STRUCTURE

Structure and Properties of Materials

VOLUME I	STRUCTURE
	William G. Moffatt George W. Pearsall John Wulff
VOLUME II	THERMODYNAMICS OF STRUCTURE
	Jere H. Brophy Robert M. Rose John Wulff
VOLUME III	MECHANICAL BEHAVIOR
	Wayne Hayden William G. Moffatt John Wulff
VOLUME IV	ELECTRONIC PROPERTIES
	Leander Pease Robert M. Rose John Wulff

STRUCTURE

William G. Moffatt

George W. Pearsall

John Wulff

JOHN WILEY & SONS, INC., *New York · London · Sydney*

10

ISBN 0 471 61265 0
Library of Congress Catalog Card Number: 64-23854
Printed in the United States of America

Preface

The four brief volumes in this series were designed as a text for a two-semester introductory course in materials for engineering and science majors at the sophomore-junior level. Some curricula provide only one semester for a materials course. We have found that under such circumstances it is convenient to use Volumes I, II, and parts of III for aeronautical, chemical, civil, marine, and mechanical engineers. Similarly, parts of Volumes I, II, and IV form the basis for a single-semester course for electrical engineering and science majors.

The four volumes grew from sets of notes written for service courses during the last decade. In rewriting these for publication, we have endeavored to emphasize those principles which relate properties and behavior of different classes of materials to their structure and environment. In order to develop a coherent and logical presentation in as brief a context as possible, we have used problem sets at the end of each chapter to extend and illustrate particular aspects of the subject. Real materials encountered in engineering situations have been chosen as examples wherever possible. Problem and laboratory sections to supplement lectures have aided our students considerably in applying the principles delineated in the text to a variety of materials and environments.

Many of the tables and illustrations used in the present text have been borrowed from individual specialists and their publications. Our thanks are due both to the individuals responsible for the data and to the publishers. Their names are listed with the illustrations in the text. Further thanks are due to numerous colleagues who took part in teaching the same courses with us during the past ten years. Many parts of the text have been improved as a result of their constructive criticism.

Finally, we wish to acknowledge our indebtedness to the Ford Foundation and to Dr. Gordon S. Brown, Dean of Engineering at M.I.T., who early supported our efforts to provide lecture demonstrations, laboratory experiments, and notebook editions of the present text for the use of our students.

June 1964

William G. Moffatt, General Electric Company
George W. Pearsall, Duke University
John Wulff, Massachusetts Institute of Technology

Contents

CHAPTER ONE

Electrons and Bonding

The ways in which atoms are arranged in a material are determined primarily by the strength and the directionality of the interatomic bonds. Qualitatively, we can understand why an atomic bond is strong or weak, directional or nondirectional, from a knowledge of the energies and the locations of the bonding electrons with respect to the positively charged ion cores. A pronounced lowering of electron energies results in a strong, or primary, bond; a slight lowering results in a weaker, or secondary, bond. Three limiting cases of primary bonding are defined. They are covalent, metallic and ionic bonding; one is distinguished from another by the way in which the bonding electrons are localized in space. The limiting cases of secondary bonding are not as easily separable, for all of them may be viewed in terms of weak dipole attractions.

1.1 INTRODUCTION

Since electrons obey the laws of quantum mechanics, our understanding of their behavior must be based on a knowledge of these laws. The most important aspects (for our purposes) of the fact that electrons are quantum mechanical particles are summarized below.

1. The principal feature of quantum mechanics is the quantization of the energies which an electron can have. That is, an electron can have only certain energies, and it cannot have any energy between these allowed values. If an electron changes its energy, it must make a *quantum jump* to another allowed energy level. The electron can jump to a higher energy level by absorbing radiant energy or to a lower energy level by emitting radiant energy. The

frequency ν of the energy emitted or absorbed is given by the Einstein equation

$$\Delta E = E_2 - E_1 = h\nu \tag{1.1}$$

where h is Planck's constant, E_2 is the energy after the jump and E_1 is the energy before the jump: a *negative* ΔE means that energy is *emitted.*

2. A second feature of quantum mechanics, the *Pauli exclusion principle,* states that no more than two electrons can occupy the same energy level and that these two must have opposite spins. Therefore, for any atom larger than helium, all the electrons cannot have the same low energy; all but the lowest two must occupy higher energy levels. As will be shown in the next sections, it is the quantization of electron energies together with the exclusion principle that is responsible for different bonding tendencies among the elements.

3. The *uncertainty principle* is a third feature of quantum mechanics which is particularly important in determining how we interpret the results of quantum mechanical conditions; its validity prevents us from describing the motion of electrons completely. This principle, initially formulated by Heisenberg, states that we cannot measure all the quantities describing the motion of a particle with unlimited precision. For the case of momentum and position, the limitation is of the form

$$\Delta p_x \, \Delta x \geqslant \frac{h}{2\pi} \tag{1.2}$$

for each of the cartesian coordinates, where Δp_x is the uncertainty in momentum, Δx is the uncertainty in position, and h is Planck's constant. For the case of energy and time, the limitation is

$$\Delta E \, \Delta t \geqslant \frac{h}{2\pi} \tag{1.3}$$

It is because Planck's constant is so small (6.63×10^{-27} erg-sec) that these uncertainties place severe restrictions on what we can know about small particles like electrons without limiting our observations of larger objects. As the uncertainty principle suggests, even when we find an equation describing the behavior of electrons we shall not be able to interpret the solution in as much detail as is possible for a larger object like a billiard ball or a planet.

1.2 WAVE MECHANICS AND ELECTRON BEHAVIOR

The equations of *wave mechanics* and their solutions are the mathematical tools by which we relate the consequences of quantum mechanics described above to the spatial distribution of electrons around the nucleus of an atom. The postulate that electrons behave as waves was first made in 1924 by de Broglie who predicted, by analogy to photon behavior in the wave theory of light, that the wavelength of an electron should be given by

$$\lambda = \frac{h}{p} \tag{1.4}$$

The fact that an electron has a wavelength does not necessarily mean that it is a wave; it does mean that the motion of the electron is governed by the same differential equations that describe wave motion. Schrödinger (1926) first suggested that the quantized energy levels, or quantum states, of electrons in an atom might correspond to *standing waves*[1] with different numbers of nodes (places where the wave amplitude is zero). In other words, the electrons might obey an equation similar to that of a vibrating string clamped at both ends,

$$\frac{d^2u}{dx^2} + \frac{4\pi^2}{\lambda^2}u = 0 \tag{1.5}$$

where u is the maximum transverse displacement, or amplitude, of a segment of string dx, x is the distance along the string, and λ is the wavelength. In fact the one-dimensional form of Schrödinger's standing wave equation for an electron can be obtained from Equation 1.5 by substituting the wave function ψ for the amplitude u, the de Broglie wavelength, $h/(mv) = h/p$, for λ, and $E - V$ for the kinetic energy $\frac{1}{2}mv^2$. The result is

$$\frac{d^2\psi}{dx^2} + \frac{8\pi^2 m}{h^2}(E - V)\psi = 0 \tag{1.6}$$

where E represents total energy and V represents potential energy. Of course, the three-dimensional form of this equation

[1] A *standing wave,* which does not change position with time, is to be distinguished from a *traveling wave,* which does change position.

$$\frac{\partial^2\psi}{\partial x^2} + \frac{\partial^2\psi}{\partial y^2} + \frac{\partial^2\psi}{\partial z^2} + \frac{8\pi^2 m}{h^2}(E - V)\psi = 0 \qquad (1.7)$$

or

$$\nabla^2\psi + \frac{8\pi^2 m}{h^2}(E - V)\psi = 0$$

is the one which must be used to compute electron energies and spatial distributions around the nucleus. We shall describe only the results of these computations and the physical interpretations that may be placed on them without introducing the mathematical details.[2]

1.3 ELECTRONIC STRUCTURE OF ATOMS

The solutions to the wave equation (Equation 1.7) for electron distributions around a nucleus are usually written in the form $\psi = \psi(r, \phi, \theta)$ where r, ϕ, and θ are spherical coordinates (Figure 1.1). These solutions resemble the solutions of the vibrating string equation in the sense that they are quantized and contain an integral number of nodes, but the physical interpretation of a solution for an electron is fundamentally different from that for a string. In the first place, the wave function ψ cannot be interpreted simply as an amplitude because the uncertainty principle states that we cannot know the position of the electron precisely enough to specify it in this way. We therefore use the interpretation given in 1926 by Born: $\psi^2\,dV$ is the probability[3] that the electron will be

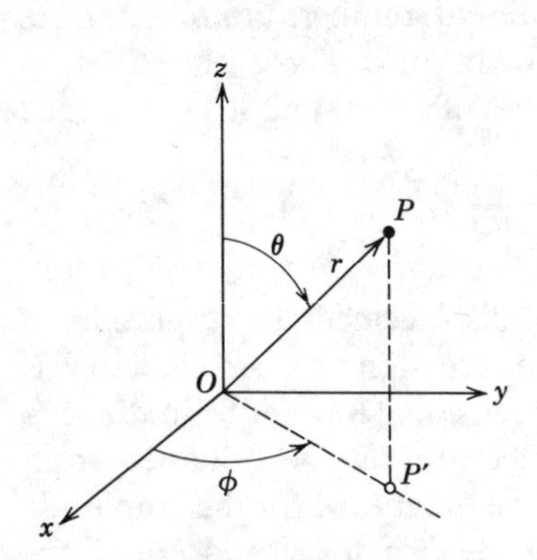

Figure 1.1 The spherical coordinate system used in describing electron distributions around the nucleus. The vector $\overline{OP'}$ is the projection of the vector $\overline{OP}$ onto the xy plane.

[2] The details of solving Schrödinger's equation for selected problems may be found in *Modern Physics* (John Wiley, 1963) by R. L. Sproull.

[3] The wave function ψ is not restricted to being a real number and may be complex. When it is complex ψ^2 should be written as $\psi\psi^*$, but we shall use the ψ^2 representation.

found in the volume dV, or for a spherically symmetric function $4\pi r^2\psi^2\, dr$ is the probability that the electron will be found between r and $r + dr$ (see Problem 1.2). This interpretation is reasonable, at least by analogy, since we know that for light the intensity (and, therefore, the probability of finding photons) is proportional to the square of the light wave amplitude. Another feature of the wave function ψ which distinguishes it from the amplitude of a string is that it is a three-dimensional function, and, generally, its nodes (places where $\psi = 0$) will be surfaces rather than points. The locations of these nodes around the nucleus indicate where the probability of finding an electron is zero. Fortunately, the wave function can be expressed as the product of three factors, each a function of only one variable:

$$\psi = R(r)\Phi(\phi)\Theta(\theta) \tag{1.8}$$

permitting us to classify nodes as one of three kinds: *spherical* nodes (nodes in the R function), *planar* nodes (nodes in the Φ function), and *conical* nodes (nodes in the Θ function).

The greater the number of nodes in the wave function of an electron, the higher its energy. A reasonable approximation of the energy of an electron in an allowed state, moving around a single positive charge, is given by Bohr's equation

$$E = \frac{-2\pi^2 me^4}{n^2h^2} = \frac{-13.6}{n^2}\,\text{eV} \qquad (\text{where } n = 1, 2, 3, \ldots) \tag{1.9}$$

where m is the mass of the electron, e is its charge, and n is an integer called the *principal quantum number.*

The total number of nodes (surfaces on which $\psi = 0$) is equal to n, and the *ground state,* or lowest energy state, will be that with one node, which is located at $r = \infty$. Since the probability of finding an electron between r and $r + dr$ in a spherically symmetric state is $4\pi r^2\psi^2\, dr$, this probability will be zero at $r = 0$ as well as on any surface where $\psi = 0$. The next permissible energy state has $n = 2$ and a node between $r = 0$ and $r = \infty$. The ψ curves and the probability curves for these two states are spherically symmetrical and are shown in Figure 1.2. As discussed below, the distributions are not always this symmetrical for higher values of n because planar and conical nodes become possible.

In contrast to the solutions of the one-dimensional vibrating

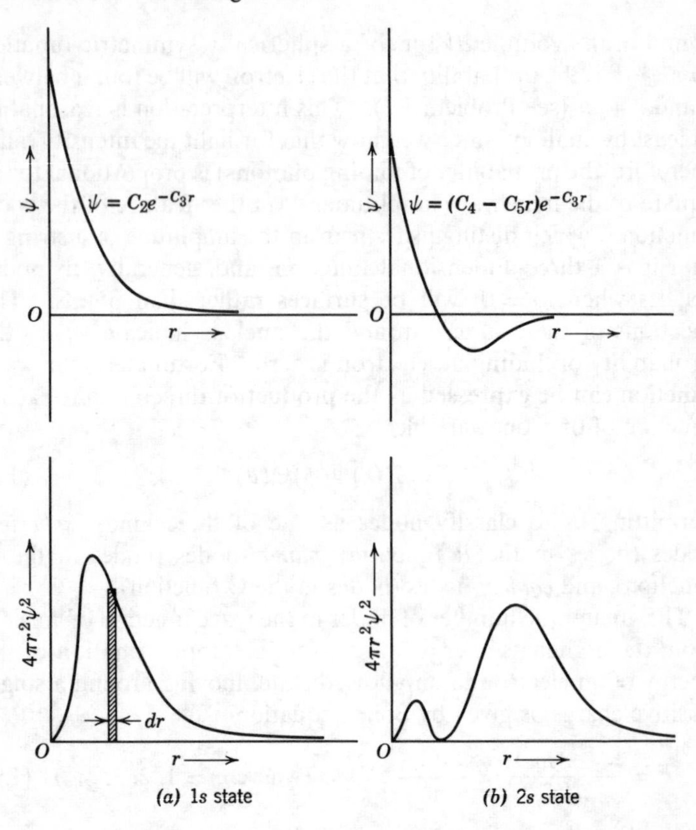

Figure 1.2 Qualitative curves of ψ and of $4\pi r^2\psi^2\,dr$ versus r for single electrons in the 1s ($n = 1$) state and the 2s ($n = 2$, $l = 0$) state. The cross-hatched strip in (a) has the area $(4\pi r^2\psi^2)(dr)$ and is thus the probability of finding the electron between r and $r + dr$. The values of the constants, C_2 to C_5, are determined by the boundary conditions imposed on the wave equation.

string problem, which are characterized by one set of quantum numbers (the number of nodes along the string), the solution to the three-dimensional wave equation must have three sets of quantum numbers, all integers. They are usually represented by the letters n, l, and m, each of which can have only certain values (Table 1.1). As described above, the principal quantum number n may be taken as a measure of how the total energy is quantized and may have values 1, 2, 3 . . . The quantum number l describes

Table 1.1 Permissible Values of the Three Quantum Numbers

QUANTUM NUMBER	PERMISSIBLE VALUES
n	1, 2, 3, 4, 5, 6, . . . (n)
l	0(s), 1(p), 2(d), 3(f), . . . ($n - 1$)
m	$+l, \ldots, +1, 0, -1, \ldots, -l$

the way in which the angular momentum of the electron is quantized and has a small effect on the energy; l can have only the values 0, 1, . . . , ($n - 1$). The l states are usually indicated by letters to avoid confusion with the values of n: s is used for ($l = 0$); p for ($l = 1$); d for ($l = 2$); and f for ($l = 3$).[4] The quantum number m is a measure of the angle between the electron's angular momentum vector and an applied magnetic field; m can have only integral values from $+l$ to $-l$, including zero. Each different combination of n, l, and m corresponds to a unique quantum state called an *orbital.* According to the Pauli exclusion principle, each orbital can contain no more than two electrons, and these two must have opposite spin (usually designated by a fourth quantum number m_s, which can have values of only $+\frac{1}{2}$ or $-\frac{1}{2}$). For cataloging the elements (see Appendix IB) it is sufficient to designate only the values of n and l and the number of electrons in each l state. For example, $(1s)^1$ represents hydrogen, $(1s)^2$ represents helium, $(1s)^2(2s)^2(2p)^4$ represents oxygen, $(1s)^2(2s)^2(2p)^6(3s)^2(3p)^2$ represents silicon, etc. Note that the permissible values of m (Table 1.1) and m_s determine how many electrons are required to fill each l state.

Whether a bond is directional or not depends on the spatial symmetry of these orbitals around the nucleus. The quantum number l gives the number of planar and conical (or nonspherical) nodes passing through the origin and, therefore, only orbitals with $l = 0$ have a spherically symmetric probability function,[5] for example, the ($1s$) and ($2s$) orbitals represented in Figure 1.2. The

[4] The letters s, p, d, and f are carry-overs from early spectrographic work and describe those states which give rise to the *s*harp, *p*rincipal, *d*iffuse, and *f*undamental spectral lines.

[5] The fact that $l = 0$ does not mean that the electron is at rest but only that its motion is as probable in any one direction as in any other.

nonspherically symmetric p orbitals are more difficult to visualize but can be illustrated by taking the *radial probability factor* R^2 and the *angular dependent probability factor* ($\Phi^2\Theta^2$) from the equation $\psi^2\, dV = (R^2\Phi^2\Theta^2)\, dV$ and considering each individually. The angular dependent factor is plotted in Figure 1.3 for the three $2p$ orbitals often encountered in solids. Figure 1.4 shows schematically the multiplication of the angular dependent factor by R^2 to obtain ψ^2, and Figure 1.5 shows the variation of ψ^2 with position in the xy plane for one of the $2p$ orbitals.[6] Although each of these three p orbitals is nonspherically symmetric, the sum of all three gives a spherically symmetric distribution of ψ^2. Likewise, both the sum of the five d orbitals and the sum of the seven f orbitals are spherically symmetric. Therefore a full *shell* of electrons (all with the same value of n) presents a spherically symmetric charge distribution around the nucleus.

1.4 THE HYDROGEN MOLECULE

The simplest example of how a knowledge of electron distributions and energies helps to explain why two atoms bond together

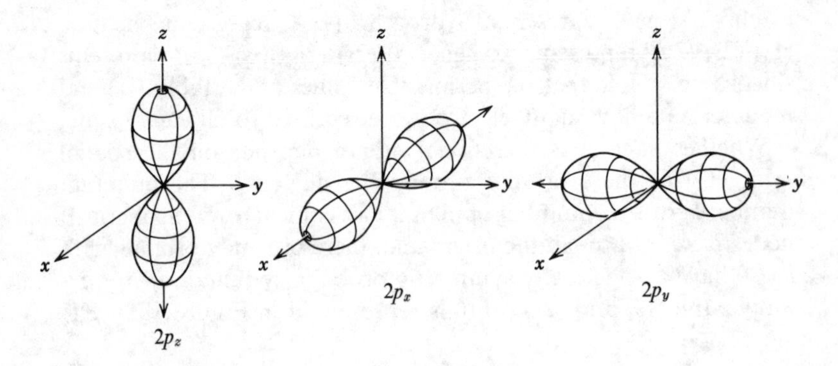

Figure 1.3 The angular dependent probability factors for the three $2p$ states often used in atomic bonding.

[6] For an indication of how ψ^2 varies with position when $l > 1$, that is, for d and f orbitals, the student is referred to *Atomic Theory for Students of Metallurgy*, Chapter XII (The Institute of Metals, 1955) by W. Hume-Rothery.

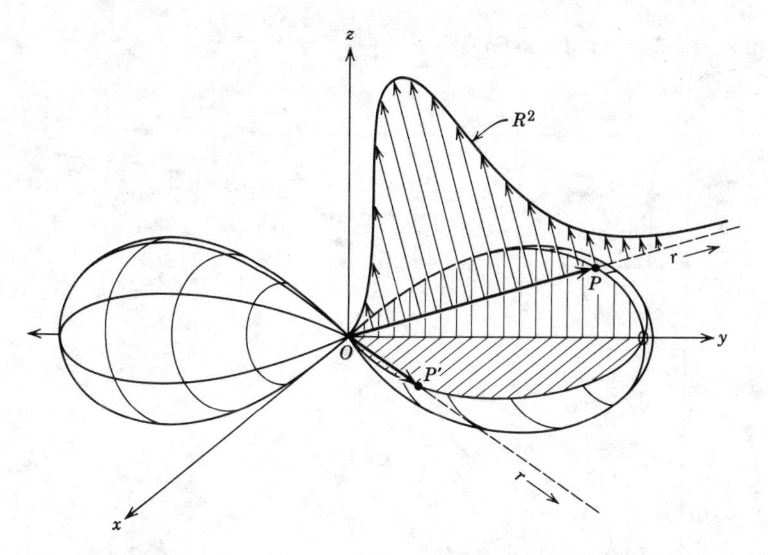

Figure 1.4 The geometry of the electron probability expression for one of the 2p states. The radius vector $\overline{OP}$ gives the magnitude of $\Phi^2\Theta^2$ which must be multiplied by the radial probability factor R^2 to give ψ^2, the probability per unit volume of finding the electron at any radius r along a line in the direction of $\overline{OP}$. For the direction defined by the vector $\overline{OP'}$, the probability curve will have the same shape but will be scaled down since $|\overline{OP'}| < |\overline{OP}|$.

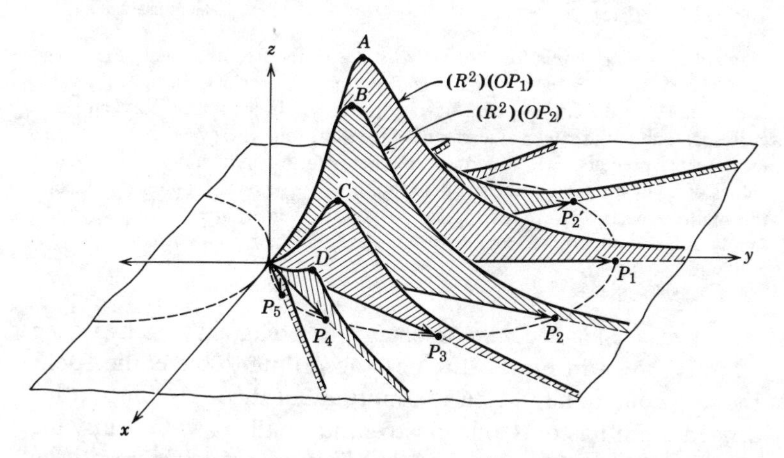

Figure 1.5 Representation of ψ^2 on the xy plane for the 2p state in Figure 1.4. The ordinate of the curves gives the probability per unit volume (ψ^2) of finding the electron at various points on the xy plane. The ratios of heights of the ordinates $A:B:C:D$ are the same as the ratios of the lengths $|\overline{OP}_1|:|\overline{OP}_2|:|\overline{OP}_3|:|\overline{OP}_4|$ because the radial probability factor R^2 is the same for all curves. The dashed curve is the angular dependent probability factor, $\Phi^2\Theta^2$.

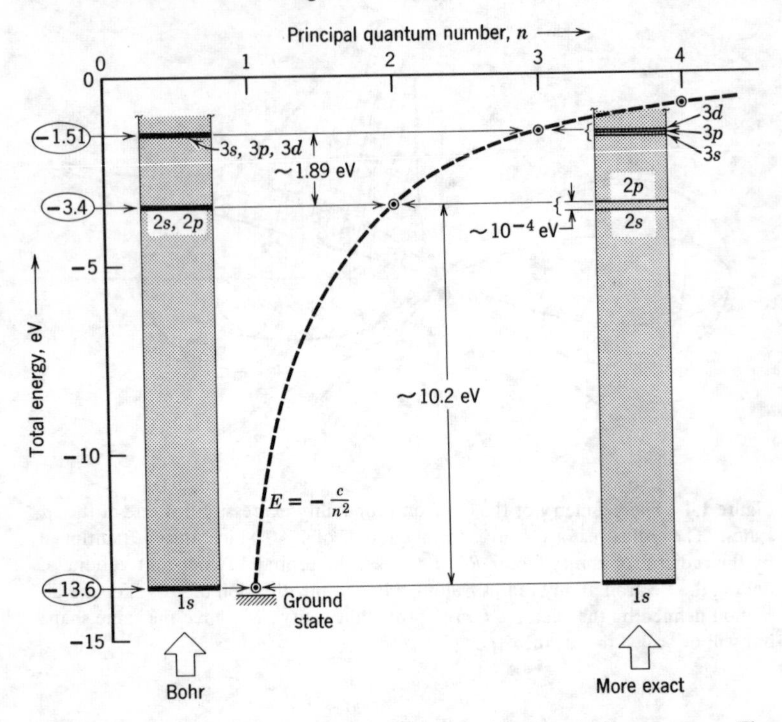

Figure 1.6 Energy levels for the excited states of the free hydrogen atom. The bold dashed curve is a representation of Equation 1.9 for which the only allowable energy states are the circled points at $n = 1, 2, 3 \ldots$. In the Bohr theory, all states corresponding to the same value of n have the same total energy, whereas the more exact theory predicts small differences in energy for different states with different values of l, but the same principal quantum number. For the more exact theory, the multiple energy levels correspond to points given by $n = 1, 2, 3 \ldots$ on a family of nearly superposed, but separate, curves for the s, p and d states.

is the hydrogen molecule. An energy level diagram for a hydrogen atom is shown in Figure 1.6, both according to the earlier Bohr theory (Equation 1.9) and the more exact theory.[7] Since a hydrogen atom has only one electron, normally it will occupy the lowest energy or 1s state. But this electron, since it is negatively

[7] For a brief summary of the developments leading to a more exact quantum theory of the atom, see *Atomic Theory for Students of Metallurgy* (The Institute of Metals, 1955) by W. Hume-Rothery.

charged, can lower its energy even more by getting closer, at the same time, to a second hydrogen nucleus (Figure 1.7). The total energy of the two atoms decreases as they move closer together, resulting in an attractive force, or *bonding force.* The lowest energy will occur when both electrons simultaneously occupy the 1*s* states of both atoms, providing the electrons have opposite spin to satisfy the exclusion principle. If the nuclei move any closer together, the electrostatic repulsion between them will tend to push them back; if they move any farther apart, the total energy of the two atoms will be increased. Therefore the molecule is stable. Of course, the presence of another atom distorts the ψ^2 distribution of the first (Figure 1.8) because there is a tendency for the electrons to spend somewhat more time between the nuclei. The bond is *covalent* when both electrons are between the nuclei, filling the overlapping 1*s* orbitals; this covalent bond is the result of both electrons being able to lower their energies by residing in the 1*s* orbitals of *two*

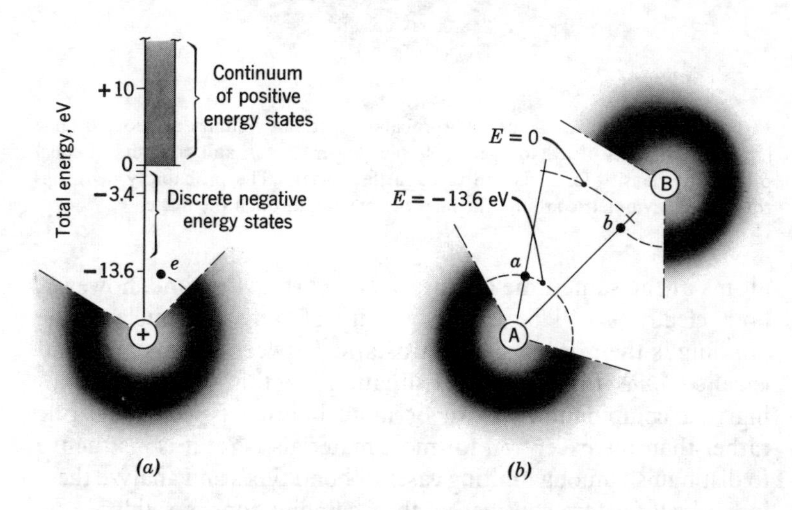

Figure 1.7 Energy levels for the hydrogen atom. (*a*) The free atom; the total energy of the electron is at a minimum with respect to its nucleus. (*b*) A pair of hydrogen atoms with electron clouds just starting to overlap. Electron *a* has minimum energy with respect to nucleus A but can lower its energy with respect to nucleus B if it moves closer to it. Simultaneously, the energy of electron *b* with respect to nucleus A will be lowered if the atoms move closer together.

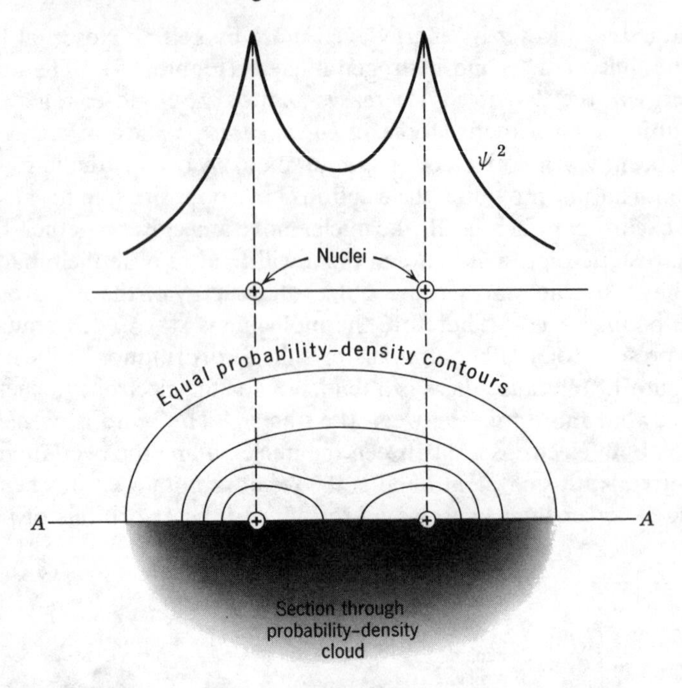

Figure 1.8 The probability density (probability per unit volume), ψ^2, along the line joining two nuclei of a hydrogen molecule (upper curve), and contours of equal probability density for a section through the nuclei. The probability density is rotationally symmetric around the line *A*-*A* passing through the nuclei.

atoms at the same time (Figure 1.9). Part of the time, however, both electrons will be in the vicinity of only one nucleus; the bonding is then primarily electrostatic, between H^+ and H^-, and is called *ionic* (Figure 1.9). A situation like this, where the bonding is a combination of two or more limiting types, is the rule rather than the exception for most materials. Yet it is instructive to distinguish among limiting cases of bond types and analyze them individually before considering their combinations in solids.

In our analysis of bond types we shall continue to use the concept of a wave function even though an accurate solution of the Schrödinger equation has been obtained only for the hydrogen molecule. Bonds between larger atoms may be rationalized by energy arguments similar to those used for the H_2 molecule. But it must be realized that the treatment is only qualitative and that only the outer, or *valence,* electrons take part in the bonding. The inner electrons are screened by the valence electrons and remain part of the *ion core.*

1.5 COVALENT BONDING

Stable covalent bonds are formed in the solid state primarily between nonmetallic atoms like nitrogen, oxygen, carbon, fluorine,

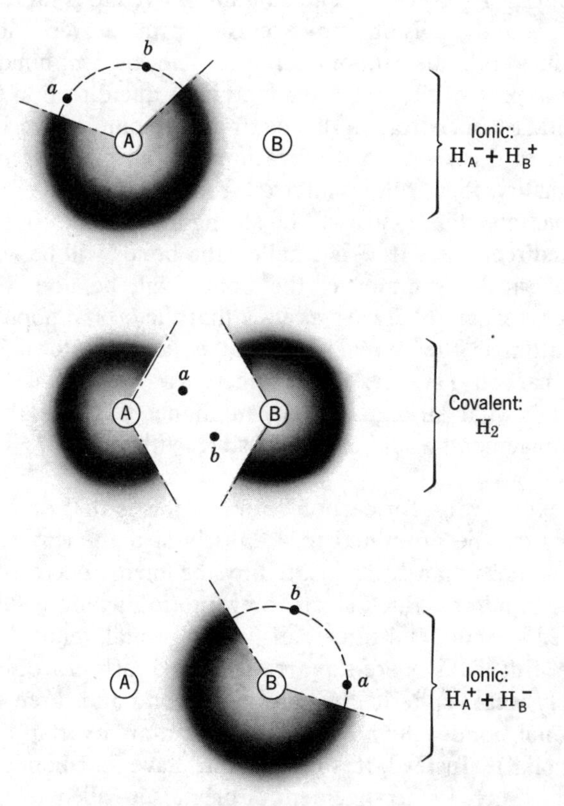

Figure 1.9 A schematic representation of ionic and covalent bonding in a hydrogen molecule. Covalent bonding results when both electrons reside in the 1*s* orbitals of both nuclei at the same time. However, there is a finite probability, at any instant, of finding both electrons associated with one nucleus. This nucleus momentarily acts like a negative ion, and the other nucleus acts like a positive ion; electrostatic attractive forces then help to hold the two "ions" together.

and chlorine. Other elements like silicon, germanium, arsenic, and selenium form bonds that are partly covalent, partly metallic; the transition metals (columns IIIB–VIIB in the Periodic Table, Appendix IA) are also thought to have a certain amount of covalent character to their bonds. A prerequisite for strong covalent bonding is that each atom have at least one half-filled orbital. Only then can the energy be lowered substantially by having each bonding electron in the orbitals of two atoms simultaneously. The more the bonding orbitals overlap, the more the energy is lowered, or the stronger the bond. The amount of overlap is limited either by electrostatic repulsion or by exclusion-principle repulsion (completely filled orbitals cannot overlap). The covalent bond formed by overlapping orbitals also tends to lie in the direction in which the orbital is concentrated, the maximum amount of overlap being obtained in this way. An indication of both bond strength and directionality, then, can be inferred from ψ^2. If ψ^2 is large in certain directions, the bonds will be strong and will be concentrated in these directions. If ψ^2 is smaller, the bonds will be weaker; if it is spherically symmetric, the bonds will be nondirectional. However, we cannot always assume that the orbital population of a free atom is the same as that of a bonded atom. For example, carbon $(1s)^2(2s)^2(2p)^2$ sometimes forms four covalent bonds of equal strength (as in methane, CH_4, or diamond); this arrangement would be impossible with only two half-filled orbitals.

The explanation for carbon's four bonds is that one of the $2s$ electrons can be promoted to a $2p$ orbital if the expenditure of energy is more than compensated by the energy decrease accompanying bonding. Such electron promotion would result in one half-filled s orbital and three half-filled p orbitals, enough for four bonds. But if this were the arrangement, carbon would have a relatively weak, spherically symmetric bond and three stronger, directional bonds (the p orbitals permit more overlap than does the s orbital). Instead, it is observed to have four bonds of equal strength. Such a rearrangement of orbitals is called *hybridization*, and, in this case, the four resulting orbitals are called $(sp)^3$ hybrids. Mathematically, the hybrid orbitals are simply an equivalent set of solutions to the time-independent wave equation for $n = 2$. Each hybrid orbital has a larger maximum value of ψ^2 and, therefore, can overlap other orbitals to a greater extent, resulting in even lower bond energies than for overlapping *p* orbitals.[8]

The explanation of the recent discovery that gases which formerly were called inert (Xe, Kr) will form stable bonds with fluorine is similar to the explanation for carbon, except in these cases the electrons are promoted from *p* to *d* levels rather than from *s* to *p* levels.

1.6 METALLIC BONDING

The majority of elements are metals. In contrast to the nonmetallic elements, they are good conductors of heat and electricity; they are all opaque and lustrous; and they can be permanently deformed. The explanation for these unique properties lies in the nature of the metallic bond. Although metallic bonding, like covalent bonding, has its origins in the lowering of an electron's energy when it is close to more than one nucleus, pronounced differences exist between the two. For example, a metallic bond can exist only among a large aggregate of atoms, while a covalent bond can occur between as few as two atoms. Also, metallic bonds are nondirectional.

The distinguishing feature of metal atoms which accounts for their different bonding characteristics is the looseness with which their valence electrons are held. The wave functions of these electrons in free atoms are spread out much more in space than those of nonmetal valence electrons with the same principal quantum number, because the metal nuclei have fewer positive charges to attract the electrons. Typically, the valence electron ψ^2 functions are so spread out that the mean electron radius in a free atom is larger than the interatomic distance in a solid metal. In a solid this means that the valence electrons are always closer to one or another nucleus than they are in a free atom and, therefore, that their potential energies are lowered in the solid. In addition, the kinetic energy of a metallic valence electron is lowered in the solid state because the ψ^2 function is extended more in space. This

[8] For the mathematical details of deriving hybrid solutions to the wave equation, see *The Nature of the Chemical Bond,* Chapter 4 (Cornell University Press, 1960) by L. Pauling.

spatial extension of the wave function may be treated approximately as an increased wavelength, which results in a decrease in kinetic energy (see Problem 1.17). It is this decrease in both potential and kinetic energies of valence electrons that is responsible for metallic bonding. Since each valence electron is not localized between only two ion cores, as in covalent bonding, metallic bonding is nondirectional, and the electrons are more or less free to travel through the solid. The diffuse nature of metallic bonding is responsible for the easy deformability of metals and is sometimes described in terms of an electron "gas" which holds the positive ion cores together. Another representation, which relates more directly to bonding orbitals, is that of a time-averaged, fluctuating covalent bond (Figure 1.10).

In general, the fewer valence electrons an atom has and the more loosely they are held, the more metallic the bonding. These elements, like sodium, potassium, copper, silver, and gold, have high electrical and thermal conductivities because their valence electrons are so mobile. They are opaque because these "free" electrons absorb energy from the light photons, and they have high reflectivity because the electrons re-emit this energy as they fall back to lower energy levels. As the number of valence electrons and the tightness with which they are held to the nucleus increase, they become more localized in space, increasing the covalent nature of the bonding. The transition metals (metal atoms with incomplete *d* shells, such as iron, nickel, tungsten, and titanium) have a significant fraction of covalent bonding, involving hybridized inner shell electron orbitals; this accounts in part for their high melting points. The competition between covalent and metallic bonding is particularly evident in the fourth column of the periodic table. Diamond exhibits almost pure covalent bonding; silicon and germanium are more metallic; tin actually exists in two modifications, one mostly covalent and the other mostly metallic; and lead is mostly metallic.

1.7 IONIC BONDING

An ionic bond results from an electrostatic attraction between positive and negative ions which are derived from the free atoms by the loss or gain of electrons. Electronegative atoms (nonmetallic atoms which tend to acquire electrons and become negative ions) are those which have only a few half-empty p orbitals; they can attract an external electron into one of the half-empty orbitals by making it possible for the electron to lower its energy with respect to that nucleus. Electropositive atoms (metal atoms which tend to lose electrons and become positive ions) are those which have one or more loosely held electrons in a higher energy level lying above a filled electron shell.

If an electropositive and an electronegative free atom are brought close enough together that they become two ions of opposite charge, the potential energy V of the ion pair will become more negative as the radial distance of separation r decreases:

$$V_{\text{attractive}} = \frac{-\text{constant}}{r} \tag{1.10}$$

As the distance of separation decreases, though, the electron clouds of the two ions will start to overlap. At this stage the Pauli exclusion principle requires that some electrons be promoted to higher energy levels. Work must then be done *on* the ions to force them closer together, and the amount of work necessary is inversely proportional to some power, m, of the distance between ion centers. The total potential energy of the two ions will be of the form

$$V_{\text{total}} = \frac{-\text{A}}{r} + \frac{\text{B}}{r^m} + \Delta E \tag{1.11}$$

where ΔE is the energy required to form the two ions from neutral atoms. In Figure 1.11*a* if $-E_\text{A}$ is the energy of an outer s electron in the electropositive atom **A**, an amount of work $0 - (-E_\text{A}) = E_\text{A}$ must be done to remove it completely from its nucleus; E_A is called the *ionizing potential.* If it is now drawn into a half-filled p orbital of the electronegative atom **B**, it will lower its energy from zero to $-E_\text{B}$, where E_B is the *electron affinity.* The net change in the energy of this electron is therefore $\Delta E = E_\text{B} - (-E_\text{A})$ and is positive, since the ionizing potential of **A** is greater than the electron affinity of **B**. This energy required to form the ions must be more than balanced by the bonding energy of the two ions for a stable bond to be formed. It can be seen in Figure 1.11*b* that, if the bonding energy is large enough, a minimum exists in the curve of total potential energy versus distance. The minimum occurs at a distance $r = d_0$ where the attractive and

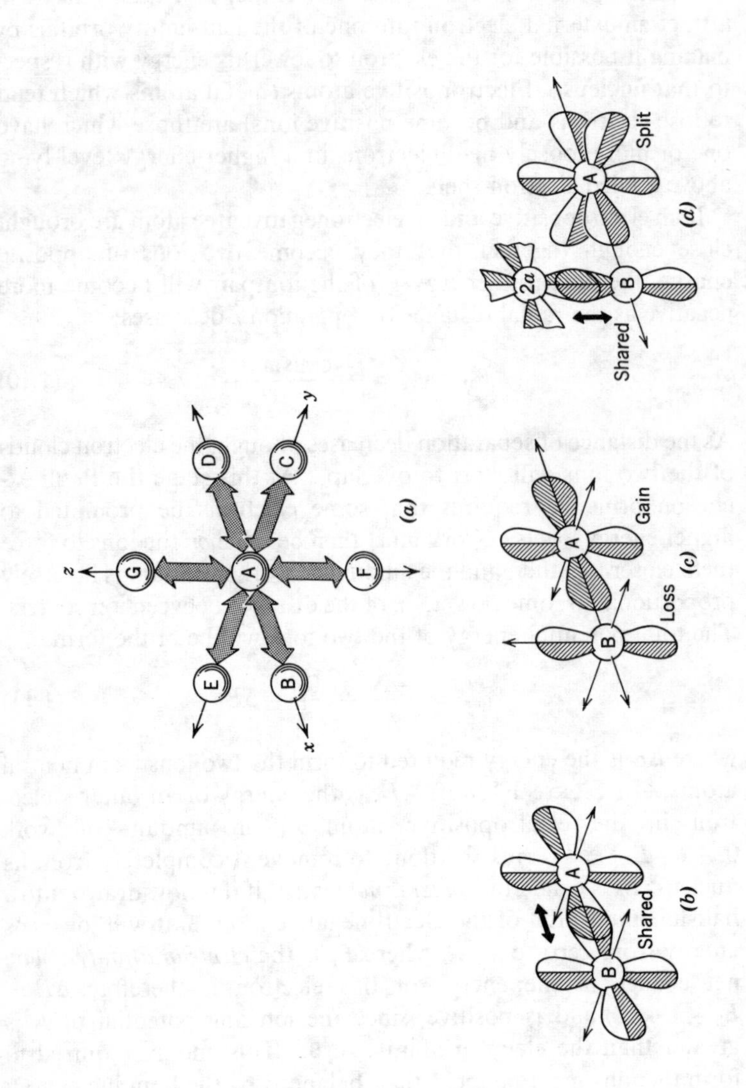

z
G
E
D
A
B
C
F
x
y
(a)
B
A
Shared
(b)
B
A
Loss
Gain
(c)
2a
B
A
Shared
Split
(d)

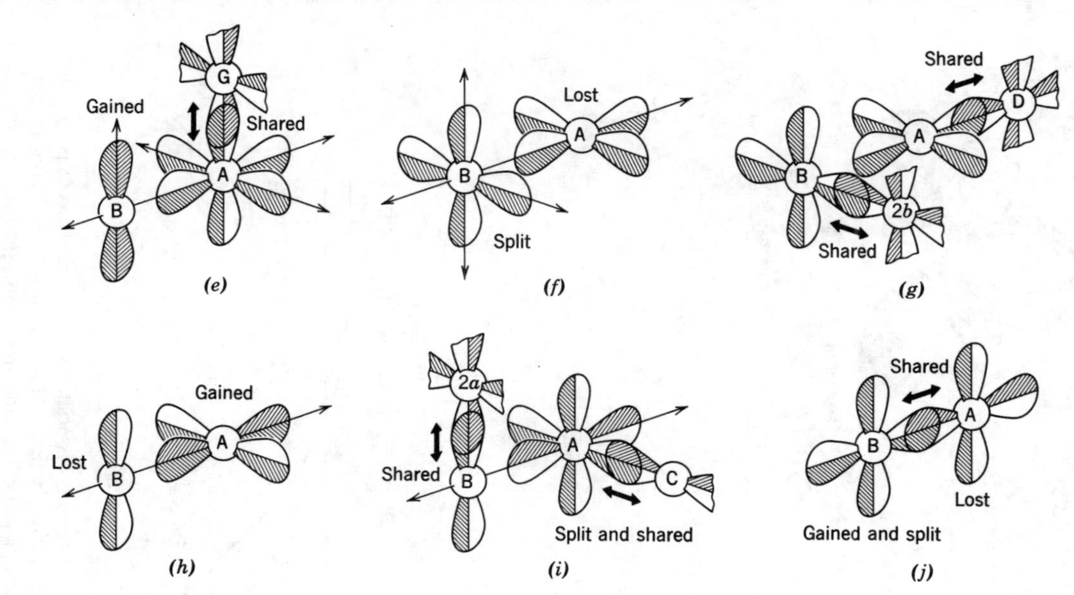

Figure 1.10 Metallic bonds viewed as time-averaged, fluctuating covalent bonds. Alternate sharing of electrons with all immediate neighbors is indicated for atoms having two p electrons, on the average. Orbitals with p electrons are indicated by stylized versions of the angular dependent wave function surfaces and are half-shaded when occupied by one electron, fully shaded when occupied by two electrons. They are not shown if vacant. Filled orbitals "split" and put one electron in a vacant orbital. Bold arrows indicate momentary covalent bonds. In (*a*) atom **A** is surrounded by six nearest neighbors; in (*b*) through (*j*) the orbitals are shown at succeeding instants of time. Atoms **A, B, C** etc. have the positions shown in (*a*); atoms 2*a* and 2*b*, not shown in (*a*), are second nearest neighbors to atom **A**. This sequence is drawn in such a way that the final arrangement is the same as the starting arrangement.

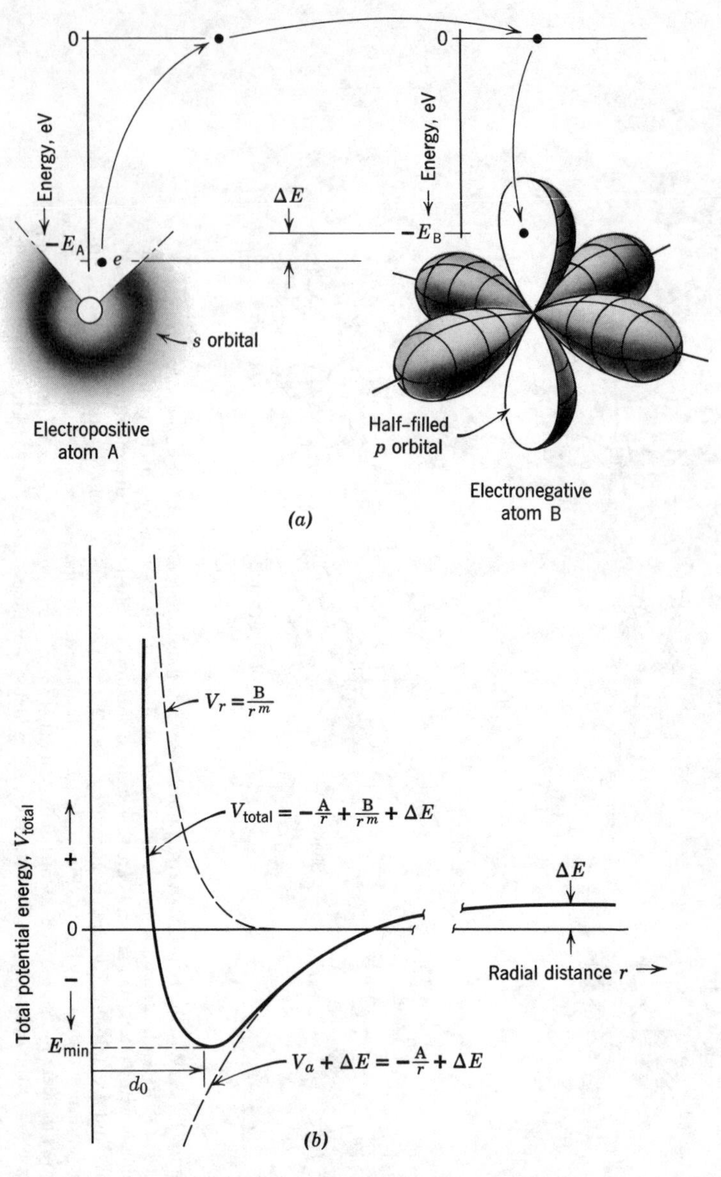

Figure 1.11 Over-all energy changes associated with the formation of a pair of ions from the free atoms and with the formation of an ionic bond. (*a*) The energy level changes in the formation of a pair of ions from free atoms. (*b*) The total potential of an ion pair as a function of radial distance from one ion. The distance $|d_0|$ is the equilibrium distance between the two ions.

repulsive forces just balance each other (see Problem 1.18); and departure from this equilibrium spacing d_0 increases the potential energy and thus produces a restoring force.

Unlike the covalent bond, the ionic bond is nondirectional. Each positive ion attracts all neighboring negative ions and vice versa, so that in a large aggregate each ion tends to be surrounded by as many ions of opposite charge as can touch it simultaneously. For example, a sodium ion attracts as many chlorine ions as will fit around it in NaCl, and lithium behaves similarly in LiF. The actual number of ions which can be accommodated is dictated first by geometric factors and second by the necessity of preserving electrical neutrality in the solid. The geometric restrictions arise because ions behave approximately as spheres, and, while many small spheres can touch a large sphere simultaneously, only a few large spheres can touch a small sphere simultaneously. This feature of ionic and atomic packing will be discussed in Chapter 2.

Whether a chemical compound will form primarily by ionic or covalent bonding depends on which mechanism will lower the total energy more. In general, the more electropositive the metal and the more electronegative the nonmetal, the greater the ionic contribution to bonding: LiF is almost completely ionic; MgO has a little covalent character to its bonds; and SiO_2 is about half ionic, half covalent.

1.8 SECONDARY BONDS

Besides the three *primary bonds* (covalent, metallic, ionic) various *secondary bonds* exist in solids. They are secondary in the sense of being relatively weak in comparison with the covalent, metallic, and ionic bonds and result from the electrostatic attraction of dipoles.

Covalently bonded atoms often produce molecules that behave as *permanent dipoles.* For example, in the water molecule, oxygen shares two half-filled p orbitals with electrons from two hydrogen atoms. Because the electrons shared between the oxygen and hydrogen atoms tend to spend most of their time between the atoms, the oxygen atom tends to act as the negative end of an electric dipole and the hydrogen atoms act as positive ends. Each of the positive ends then attaches to the negative end of another water molecule and bonds the molecules together as illustrated in Figure 1.12*a*. Another example of this molecular polarization is illustrated in Figure 1.12*b*. Here a covalent bond between hydrogen and flourine results in the formation of a dipole; and the

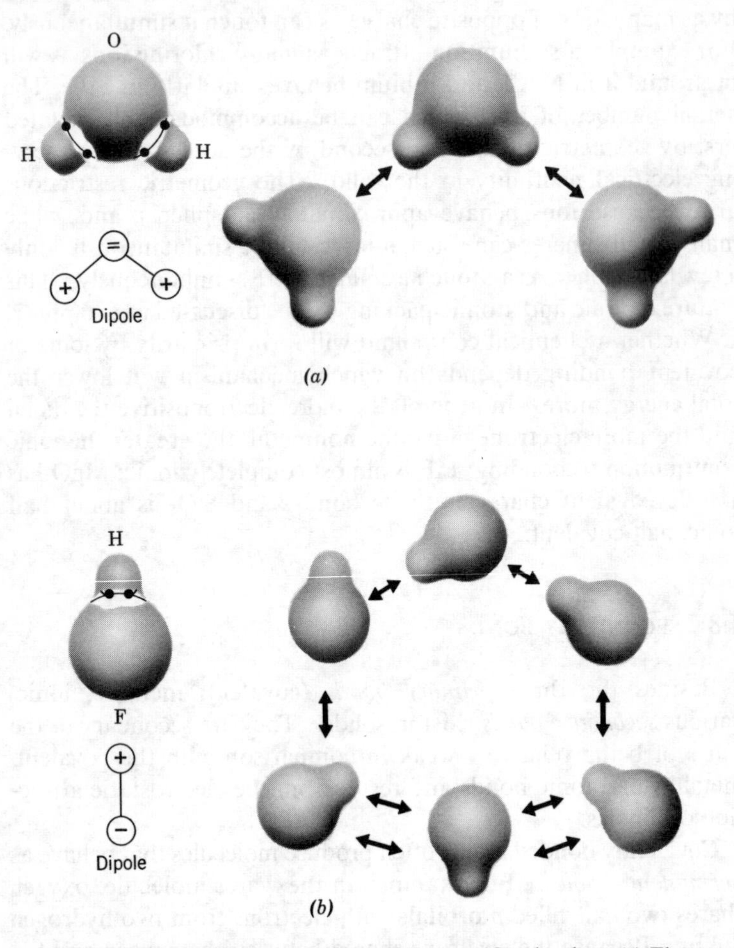

Figure 1.12 Dipole forces originating in assymetry of molecules. (*a*) The water molecule, showing the dipole and the direction of attraction between the molecules. (*b*) The hydrogen fluoride molecule, showing the dipole and the direction of attraction between molecules.

dipoles attract each other. Dipole bonds in which hydrogen is the positive end of the dipole are also called *hydrogen bonds* and can be moderately strong because the ion core of hydrogen is so small. Hydrogen bonds only form between the most electronegative atoms because these are the atoms that produce the strongest dipoles.

Permanent dipole bonds like the hydrogen bond are directional bonds, but other dipole bonds are not. These others are *fluctuating dipole bonds,* more commonly called *van der Waals bonds.* A van der Waals bond arises because there are at any one time a few more electrons on one side of the nucleus than on the other side; the centers of positive and negative charge do not coincide at this moment and thus a weak dipole is produced. A force then exists between opposite ends of dipoles in adjacent atoms and tends to draw them together. The bond produced by these fluctuating dipoles is nondirectional and is typically an order of magnitude weaker than the hydrogen bond. This force is one of the reasons, first considered by van der Waals, that real gases deviate from the perfect gas law. It is also the force that allows inert gas atoms to condense at low temperatures.

Although these secondary bonds are weak compared to covalent, metallic, and ionic bonds, they are often strong enough to determine the final arrangements of groups of atoms in solids. They are especially important in determining the structure and some of the properties of many polymers.

DEFINITIONS

Bonding Force: The force that holds two atoms together; it results from a decrease in energy as two atoms are brought closer to one another.

Covalent Bond: A primary bond arising from the reduction in energy associated with overlapping half-filled orbitals of two atoms.

Electron Probability Function: $\psi^2 \, dV$, the probability of finding an electron in the volume element, dV.

Electron Shell: A group of electrons having the same principal quantum number, n; often designated by capital letters, as K shell for $n = 1$, L shell for $n = 2$, etc.

Equilibrium Distance: The interatomic distance at which the force of attraction equals the force of repulsion between two atoms.

Ground State: The quantum state with lowest energy.

Hybridization: A rearrangement of orbitals, corresponding to replacing one set of solutions to the time-independent Schrödinger equation (1.7) by an equivalent but different set.

Hydrogen Bond: A secondary bond arising from dipole attractions where hydrogen is the positive end of the dipole.

Ion Core: An atom without its valence electron or electrons.

Ionic Bond: A primary bond arising from the electrostatic attraction between two oppositely charged ions.

Metallic Bond: A primary bond arising from the increased spatial extension of the valence electron wave functions when an aggregate of metal atoms are brought close together.

Node: A surface along which the wave function is zero, and therefore along which the probability of finding an electron is zero.

Orbital: The quantum-mechanical description of the state of an electron, including its energy and its time-averaged spatial distribution.

Pauli Exclusion Principle: The statement that no more than two electrons can occupy the same orbital, and that these two must have opposite spins.

Permanent Dipole Bond: A secondary bond arising from the attraction between dipoles, the oppositely charged ends of which are electronegative and electropositive atoms.

Quantum Jump: A change in energy from one *allowed* value to another.

Quantum Mechanics: A branch of physics in which the systems studied have only discrete values of energy, separated by forbidden regions; it is distinguished from continuum mechanics in which a continuum of energies is assumed possible.

Quantum Numbers: A set of integers (n, l, and m) representing the discrete solutions to the time-independent Schrödinger wave equation; permissible values of each are given in Table 1.1. A spin quantum number, m_s, may be included to indicate the spin of the electron ($\pm\frac{1}{2}$).

Uncertainty Principle: The statement that simultaneous measurements of pairs of conjugate variables, such as momentum and position, or energy and time, are subject to a specific indeterminacy; also called the indeterminacy principle.

Valence Electrons: Outer shell electrons which take part in bonding.

van der Waals Bond: A secondary bond arising from the fluctuating-dipole nature of an atom with all occupied electron shells filled.

Wave Function: ψ, the solution to the time-independent Schrödinger wave equation and its associated boundary conditions.

Wave Mechanics: A branch of mathematical physics involving the statements and solutions of differential equations describing wave behavior.

Numerical Constants and Conversion Factors

electronic charge, e:	4.8×10^{-10} statcoulomb
electronic rest mass, m_0:	9.1×10^{-28} gram
Planck's constant, h:	6.625×10^{-27} erg-sec

speed of light, c:	3×10^{10} cm/sec
1 electron volt, eV:	1.6×10^{-12} erg
1 Angstrom unit, Å:	10^{-8} cm
1 erg:	1 dyne cm
1 dyne:	1 gram cm/sec^2
1 dyne cm^2:	1 statcoulomb^2

BIBLIOGRAPHY

INTRODUCTORY REFERENCES:

F. O. Rice and E. Teller, *The Structure of Matter,* John Wiley and Sons, N. Y. (1949), Chapters 1–8.

R. T. Sanderson, *Principles of Chemistry,* John Wiley and Sons, N. Y. (1963), Chapters 4 and 9.

SUPPLEMENTARY REFERENCES:

D. H. Andrews and R. J. Kokes, *Fundamental Chemistry,* John Wiley and Sons, N. Y. (1962), Chapters 4, 6, and 7.

W. Hume-Rothery, *Atomic Theory for Students of Metallurgy,* Institute of Metals, London (1955), Parts I–III.

R. L. Sproull, *Modern Physics,* John Wiley and Sons, N. Y. (1963), Chapters 6–8.

MORE ADVANCED TEXTS:

C. A. Coulson, *Valence,* Oxford University Press (1961).

W. Heitler, *Elementary Wave Mechanics,* Clarendon Press, Oxford (1956).

L. Pauling, *The Nature of the Chemical Bond,* Cornell University Press, Ithaca (1960).

PROBLEMS

1.1 What would be some of the consequences if Planck's constant were 6.63×10^{-27} kcal-sec instead of 6.63×10^{-27} erg-sec? 1 calorie = 4.10×10^7 ergs.

1.2 Integrate $dV = r^2 \sin \theta \, dr \, d\theta \, d\phi$ over the range $0 - \pi$ in θ and $0 - 2\pi$ in ϕ to show that $\psi^2 \, dV = 4\pi r^2 \psi^2 \, dr$ when coordinates are changed from cartesian to spherical, where ψ is a function of r only.

1.3 Calculate the wavelength of radiation emitted when an electron in the hydrogen atom jumps from an excited $n = 2$ state to the $n = 1$ state.

1.4 A single electron moving in the field of a nucleus having a charge Z times the elementary charge has a total energy:

$$E = \frac{-2\pi^2 m Z^2 e^4}{n^2 h^2} = \frac{-13.6 Z^2}{n^2} \qquad \text{(cf. Equation 1.9)}$$

The energy of the 1s levels in neutral free atoms of He and Li are about

-24 and -65 eV respectively. Calculate the energies for the single electron in singly ionized He and doubly ionized Li. What is a likely reason for the discrepancy between the given and calculated values?

1.5 X-ray spectra indicate that the innermost electrons in the heavier elements are not strongly affected by the presence of the outer electrons. The following table lists very approximately energy values for the $1s$, $2s$, and $3s$ electrons in the free atoms of several elements.

	Mg	Cu	Mo	Xe	Hg	U
$1s$	-1300	-9000	$-20{,}000$	$-35{,}000$	$-85{,}000$	$-115{,}000$
$2s$	-85	-1100	-2800	-5500	$-15{,}000$	$-23{,}000$
$3s$	-7.5	-110	-500	-1150	-3500	-5700

Use the equation given in Problem 1.4 to compute values for the $1s$, $2s$, and $3s$ levels for each of the elements listed. Compare the values in the table above with those you have computed to determine for what electrons and what elements the equation is reasonably true.

1.6 The total energy of a $1s$ electron in tungsten is $-70{,}000$ eV. If a $1s$ electron is ejected by a beam of incident high-velocity electrons, it will be replaced by another electron from an outer shell, usually a $2p$ electron. The wavelength of the radiation emitted by the atom due to the $2p \rightarrow 1s$ transition is about 0.21 Å (actually, the atom emits radiation of two different but nearly equal wavelengths). Estimate the total energy of the $2p$ level in tungsten.

1.7 Write the electron configuration for each of the following elements, using only their atomic numbers and Table 1.1 as your guide: Li ($Z = 3$), Al ($Z = 13$), Ti ($Z = 22$), Fe ($Z = 26$). Compare your answers to Appendix IB.

1.8 Elements 21 through 28, 39 through 46, 57 through 78 and from 89 on are known as "transition elements". Look up the electronic configurations in Appendix IB and state the characteristic common to each of these series.

1.9 Using the concepts of quantum and wave mechanics, discuss why a covalent bond tends to be strongest in the directions where ψ^2 is a maximum.

1.10 Explain how an "inert gas" like Xe can react with fluorine to form a stable compound. Why does it not react with nitrogen?

1.11 The melting points and vaporization points of the alkali metals are as follows: Li ($T_m = 186°$C, $T_v = 1336°$C); Na ($T_m = 97.5°$C, $T_v = 880°$C); K ($T_m = 62.3°$C, $T_v = 760°$C); Rb ($T_m = 38.5°$C, $T_v = 700°$C); Cs ($T_m = 28.5°$C, $T_v = 670°$C). Rationalize this sequence in terms of the electronic structures and the bonding characteristics of these elements.

1.12 What effects might you expect the electron configuration of a transition metal to have on its bonding characteristics? Compare the strength of bonding and the engineering usefulness of transition metals with the same properties of the alkali metals.

1.13 Explain why carbon atoms in diamond bond covalently, while lead atoms bond metallically, even though carbon and lead have four valence electrons each. What effects would you expect this difference in bonding to have on strength, ductility and conductivity?

1.14 For a single ionic molecule, the negative term in Equation 1.11 is the coulomb energy of attraction which, at the equilibrium spacing d_0, has the value $-A_0/d_0 = Z^2e^2/d_0$, where Z is the valency of the ions. For univalent ions, show that the expression for the total potential energy V_0 of a single molecule is

$$V_0 = \left(\frac{-e^2}{d_0}\right)\left(1 - \frac{1}{m}\right)$$

1.15 For a single molecule of univalent ions, Equation 1.11 can be written as

$$V = \frac{-e^2}{r} + \frac{B}{r^m}$$

where e is the electronic charge and the reference state is assumed to be two *ions* at infinite separation (i.e. $\Delta E = 0$). Since e is a constant, we may replace $B = Ce^2$, where C is another constant; then

$$V = \frac{-e^2}{r} + \frac{Ce^2}{r^m}$$

Obtain an expression for the equilibrium spacing of the ions as a function only of C and m.

1.16 The quantity ΔE (Equation 1.11) is the difference between the ionization potential of one atom and the electron affinity of a second atom to produce a pair of ions.

For Na and Cl:	*First Ionization Potential*	*Electron Affinity*
Na	5.13 eV	0
Cl	12.95 eV	3.8 eV

Calculation of covalent bond energy yields −58 kcal/mole for Cl—Cl and −68 kcal/mole for Na—Cl; calculation of ionic bond energy from the equation

$$V_0 = \frac{-NZ^2e^2}{d_0}\left(1 - \frac{1}{m}\right)$$

shows that Cl^+—Cl^- has a lower energy than does Na^+—Cl^-. Explain why covalent bonding exists in Cl_2 and ionic bonding exists in NaCl.

(Use a value of $m = 9$; $d_0 = 1.988$ Å for $Cl^+—Cl^-$, and $d_0 = 2.814$ Å for $Na^+—Cl^-$; N = Avogadro's number.)

1.17 Using Equation 1.4, show that if a metallic valence electron has a longer wavelength in a solid than in a free atom, it will have a lower kinetic energy.

1.18 Force is the negative derivative of potential energy with respect to radial distance: $F = -dV/dr$, and a negative force is defined as attractive, a positive force repulsive. By reference to Figure 1.11, show that work is done *by* two ions as they move toward their equilibrium spacing and that the minimum in potential energy corresponds to zero force.

1.19 Explain how the small size of the hydrogen ion core helps rationalize the fact that hydrogen bonds are stronger than other dipole bonds. Could the hydrogen bond conceivably be classified as an ionic bond? a metallic bond? a covalent bond?

1.20 Why do inert gas atoms form in condensed (solid and liquid) states at very low temperatures, but not at room temperature?

CHAPTER TWO

Atomic Packing

The *local* arrangements of atoms in a solid are usually regular and predictable even though the long-range structure may be either regular (crystalline) or irregular (glassy). The arrangements we observe depend partly on whether the bonding is directional or nondirectional. If the bonding is directional, the local atomic arrangement is determined by the bond angles and may be represented by a bonding polyhedron, the corners of which represent directions of maximum bond strength. If the bonding is nondirectional, the arrangement depends on the relative sizes of the atoms. The packing arrangement in this case may be described geometrically in terms of a coordination polyhedron, which is constructed by connecting the centers of all the neighboring atoms or ions that touch a central one. The long-range structure of a solid can then be described either in terms of atom packing or in terms of arrangements of bonding polyhedra or coordination polyhedra, depending on which description has the most physical significance.

2.1 INTRODUCTION

When trying to understand why atoms are arranged the way they are in a solid, it is convenient to divide interatomic bonds into two categories, *directional* and *nondirectional.* Covalent and permanent dipole bonds are directional; metallic, ionic, and van der Waals bonds ideally are nondirectional. Atoms bonded by directional bonds are packed in a way that satisfies the bond angles. Atoms bonded by nondirectional bonds behave in general as tightly packed spheres and obey certain geometrical rules dictated by their difference in size. Of course, it must be remembered that, although the two types are discussed separately here for convenience, bonding in actual materials is often a mixture of both.

2.2 DIRECTIONALLY BONDED ATOMS

Of the two types of directional bonds, covalent and permanent dipole, the covalent bonds are the only ones which determine *local* packing arrangements to any significant degree, dipole bonds usually linking together groups of atoms rather than individual atoms. The characteristics of the covalent bond which determine the arrangement of atoms are its discreteness and its spatial direction. Its spatial direction is, in turn, determined by the quantum state of the electrons taking part in the bond. The distribution of an *s* electron around the nucleus is spherically symmetrical and therefore has no preferred direction. However, it is discrete in that it can form only one bond to another atom. Other orbitals are spatially directed as well as discrete; the *p* orbitals, for example, are perpendicular to one another (Figure 1.3). Only half-filled orbitals can take part in bonding and each half-filled orbital can account for one covalent bond only, so the number and arrangement of bonds is dictated by the number and arrangement of half-filled orbitals.

The most stable interatomic bonds are formed between those orbitals that can overlap the most; therefore an element with *p* electrons will tend to use them for bonding rather than *s* electrons in the same shell if a choice exists. The resulting bond angles are close to 90° because this is the angle the *p* orbitals make with each other. Table 2.1 lists some experimentally measured bond angles thought to be produced by *p* electron bonding. The larger angles between bonds in H_2O and NH_3 may be accounted for to some extent by the partially ionic character of the bonds, which causes a mutual repulsion between hydrogen atoms. The deviations from 90° of the bond angles of the elements in Table 2.1 are probably produced by some hybridization of the *s* and *p* wave functions.

An extreme example of hybridized *s* and *p* wave functions occurs in carbon bonds formed in diamond and in many organic molecules, such as methane. There are four orbitals available to the

Table 2.1 Measured Bond Angles due to Covalent Bonding

ELEMENT OR COMPOUND	ANGLE DESCRIPTION	MEASURED ANGLE
P	P-P-P	99°
As	As-As-As	97
Sb	Sb-Sb-Sb	96
Bi	Bi-Bi-Bi	94
S	S-S-S	107
Se	Se-Se-Se	104
Te	Te-Te-Te	104
H_2S	H-S-H	92
H_2O	H-O-H	104
NH_3	H-N-H	107

four electrons in the valence shell of a carbon atom: one 2*s* and three 2*p* orbitals. But instead of three relatively strong bonds at 90° to one another and one weaker, spherically symmetric bond, carbon atoms often have four bonds of equal strength directed toward the corners of a regular tetrahedron, as shown in Figure 2.1.

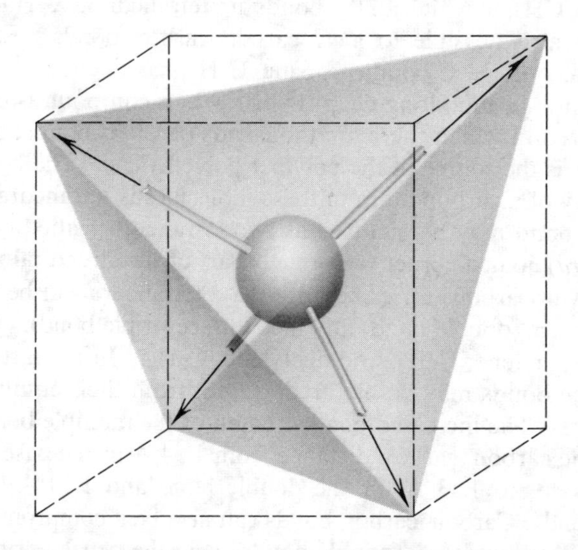

Figure 2.1 The tetrahedral directionality of the four bonds resulting from the hybridization of one *s* orbital and three *p* orbitals. The tetrahedron is shown inscribed within a reference cube.

These bonds are the result of sp^3 hybridization which, as discussed in Chapter 1, produces a set of orbitals that are mathematically equivalent to the *s* and *p* orbitals; they differ from the *s* and *p* orbitals primarily in permitting more overlap and in their spatial orientation. Silicon, germanium, and tin also have tetrahedral orbitals, as 3*s*-3*p*, 4*s*-4*p*, or 5*s*-5*p* hybridization is similar to 2*s*-2*p* hybridization. Other hybrid bond configurations are possible for elements whose bonding orbitals are different from those of the Group IV elements. For example, platinum and palladium sometimes form hybrid orbitals directed at the corners of a regular octahedron by combining two *d* orbitals with the *s* and *p* orbitals.

The fact that a covalently bonded atom forms only a certain number of discrete bonds seems almost trivial; yet it accounts both for the great variety of organic structures and for many of their unique properties. Because the bonds are discrete in number and location, many nonequivalent structures are possible, depending only on which bond is used for which atom. The variety of possibilities is illustrated by the hundreds of different compounds containing only carbon and hydrogen. The simplest hydrocarbon is methane, CH_4, in which all the bonds are tetrahedral C—H bonds. But it is also possible to have carbon–carbon bonds. Ethane, C_2H_6, has one C—C bond; propane, C_3H_8, has two C—C bonds, and so on. In fact, long chain hydrocarbon compounds can be synthesized in which there are thousands of C—C bonds; such a synthesis is the source of the polymer polyethylene. The bonding symmetry of a carbon atom in these molecules is tetrahedral, and a C—C bond may be pictured as two tetrahedra (called *bonding tetrahedra*), joined corner-to-corner. Part of the structural variety of the hydrocarbons arises because these tetrahedra can be joined edge-to-edge (double bond) and face-to-face (triple bond). Examples are ethylene, C_2H_4, and acetylene, C_2H_2. In the latter two cases, the bonds may be pictured as bent from their equilibrium directions. The increased bond strength of the multiple bonds reduces the carbon–carbon distance from 1.54 Å in the case of the single bond, to 1.33 Å for the double bond and 1.20 Å for the triple bond. Carbon–carbon bonds can also be a compromise between single bonds and double bonds, as in the benzene ring.

The capacity of a covalent bonding orbital to be completely satisfied by overlapping with only one other orbital stems from the fact that the bonding electrons are very localized in space, and therefore the bonding forces extend over very short distances. In methane, for example, the four bonds of carbon are completely satisfied by the four hydrogen atoms, and each hydrogen atom is satisfied by the carbon. Therefore methane exists only as small molecules of CH_4 weakly bonded to each other by van der Waals forces. It solidifies only at very low temperatures. In fact, methane, propane, and the other longer chain compounds in this series are called *paraffins,* which means "little affinity." The only covalently bonded materials that are solid at room temperature and above are those which possess either a three-dimensional network of covalent bonds, such as diamond and the polymer Bakelite, or such large molecules that the van der Waals forces between them are appreciable. High molecular weight polymers fall into the latter class.

2.3 NONDIRECTIONALLY BONDED ATOMS OF EQUAL SIZE

We expect nondirectionally bonded atoms like those of the metallic elements and the noble elements to solidify in as closely packed an arrangement as possible. In this way the number of bonds per unit volume is maximized, and hence the bonding energy per unit volume is minimized. As a first approximation, then, these atoms can be treated as hard spheres packed together. Geometrically, twelve is the maximum number of spheres that can be packed around a central sphere so that they all touch it simultaneously. Two such arrangements, which are nonequivalent, are observed in many metallic and noble element structures; they are the *hexagonal close-packed* (HCP) and *face-centered cubic* (FCC) arrangements shown in Figure 2.2. In each type of packing, six spheres surround a central sphere in a midplane with three more spheres above and three below. The only difference between the two is that in the HCP arrangement the three spheres above the midplane are directly above the three spheres below the midplane, and in the FCC arrangement they are in the alternate set of positions.

The designations "hexagonal close-packed" and "face-centered cubic" are derived from the geometry that results when each type

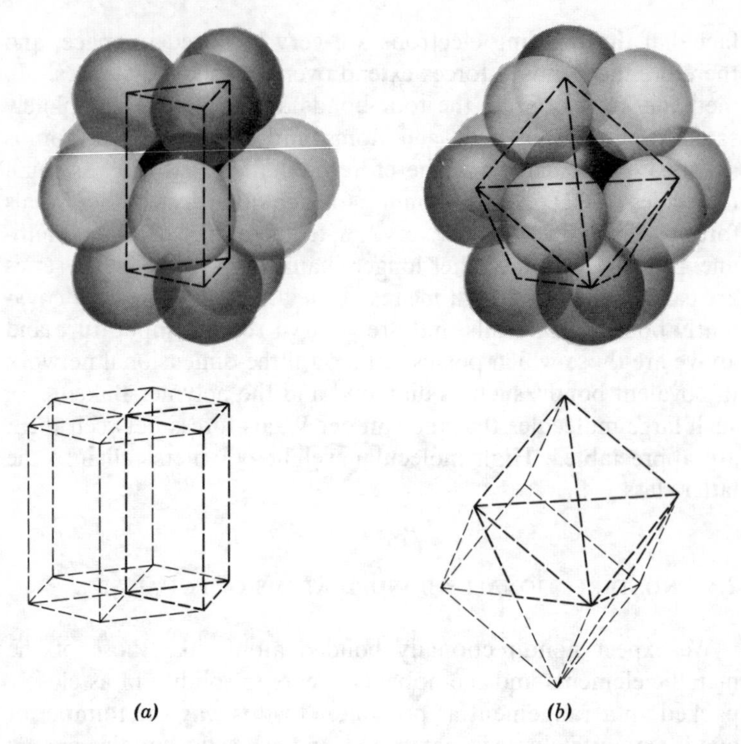

Figure 2.2 (*a*) Hexagonal close-packing of twelve atoms around a central atom. (*b*) Face-centered cubic packing of twelve atoms around a central atom. The dashed figures show the hexagonal and the cubic nature of the packing in each case. Atoms whose centers lie at the intersections of the heavy dashed lines are shaded darker than the others. Atoms whose centers lie at the intersections of the light dashed lines do not touch the central sphere and are not shown explicitly.

of packing is continued in three dimensions.[1] In Figure 2.2 the heavy dashed lines connect the centers of the darkly shaded atoms, and the light dashed lines connect atom sites that would be filled if the same packing geometry were extended by adding more atoms. The dashed figures form a *hexagonal* prism in the first case and a *cube* with an atom in the center of each face in the second case. The second type of packing is also referred to as cubic

[1] Regular arrangements of atoms that extend through space are called crystal structures and are discussed in the following chapter. The long-range structure of HCP and FCC packing will be described there.

close-packed (CCP), but we shall use the more common FCC description.

The local atomic packing arrangements that are observed among the noble and metallic elements illustrate to what extent non-directionally bonded atoms of equal size actually are close-packed. All the noble elements exhibit close-packed structures when they solidify at very low temperatures, and about two-thirds of the metals exhibit either HCP or FCC packing at room temperature. Of these two-thirds, atoms of those metals that have HCP structures pack together more like slight ellipsoids than perfect spheres, the hexagonal prism in Figure 2.2*a* being somewhat shorter or taller in proportion to its edge length than would be calculated for spherical atoms. Most of the remaining one-third that do not solidify in a close-packed structure are alkali metals (Na, K, etc.) and transition metals (Fe, Cr, W, etc.) which tend to have a *body-centered cubic* (BCC) structure like that shown in Figure 2.3. A body-centered cubic structure is not quite as closely packed as a face-centered cubic or hexagonal close-packed structure (see Problems 2.7 and 2.8) but, even so, it is a relatively low energy arrangement of atoms. Most of the alkali metals transform from BCC to FCC or HCP at very low temperatures, suggesting that one reason for their less dense packing behavior at room temperature is the effect of thermal energy.[2] The reason that most of the transition metals solidify in a BCC structure is thought to be the partly covalent and, therefore, directional nature of their bonds, as mentioned in Chapter 1.

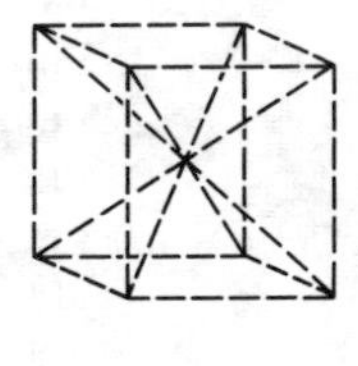

Figure 2.3 Body-centered cubic packing of atoms. The central atom has only eight neighboring atoms touching it.

[2] This subject involves the thermodynamic concept of entropy which will be discussed in Volume II, *Thermodynamics of Structure.*

2.4 NONDIRECTIONALLY BONDED ATOMS OF UNEQUAL SIZE—IONIC PACKING

Other packing arrangements of nondirectionally bonded atoms become possible if the sizes of the atoms are appreciably different. Ionic solids provide the best examples of packing determined by the relative sizes of atoms (ions) because anions and cations usually differ appreciably in size. Pronounced size differences result because a transfer of electrons from electropositive to electronegative atoms makes the resulting anions much larger than the cations. The bonding is nondirectional, as discussed in Chapter 1. The bonds are also not discrete, even though electrical neutrality must be satisfied in the long-range structure. This restriction on the long-range structure does not affect *local* packing arrangements and will be discussed in the following chapter on crystal structure.

Ligancy, or *coordination number,* is the name given to the number of ions of element A surrounding a smaller ion of element C. As shown in Table 2.2, the ligancy is a function of the difference in size of the C and A ions: the greater the size difference, the smaller the ligancy. The only values of ligancy possible in a three-dimensional, regular array are 1 (which is trivial), 2, 3, 4, 6,

Table 2.2 Ligancy Calculated as a Function of Radius Ratio

LIGANCY	RANGE OF RADIUS RATIO FOR WHICH LIGANCY IS EXPECTED TO BE STABLE	COORDINATION POLYHEDRON	PACKING
2	0 to .155	line	linear
3	.155 to .225	triangle	triangular
4	.225 to .414	tetrahedron	tetrahedral
6	.414 to .732	octahedron	octahedral
8	.732 to 1.0	cube	cubic
12[a]	1.0	Figure 2.4	HCP
12[a]	1.0	Figure 2.4	FCC

[a] Note that the limiting case of ligancy 12 is for both "ions" the same size, and the two types of packing are those already discussed for atoms that are the same size.

8, and 12.[3] Therefore, although the procedure we shall use is general, these are the only ligancies for which we shall calculate the relative sizes of ions. It is possible to calculate the range of radius ratio (ratio of cation radius to anion radius) over which each value of ligancy might be expected to be stable if the following conditions are assumed to exist:

1. The cations and anions are at their equilibrium interionic spacing (the spheres representing them are just touching).
2. The anions do not overlap each other (their centers do not approach more closely than an ion diameter).
3. Each cation tends to be surrounded by the largest possible number of anions (as the difference between ion sizes becomes smaller, higher values of ligancy can occur).

According to these conditions, a given ligancy is predicted to be stable between that radius ratio at which the anions touch each other as well as the central cation and that radius ratio at which the next highest ligancy becomes possible. The radius ratio at which the anions just touch each other as well as touching the central cation is called the *critical radius ratio* because below this ratio the cation-anion distance becomes greater than the equilibrium, interionic distance. For example, a ligancy of 4, or tetrahedral coordination, is predicted to be stable between radius ratios of .225 and .414. Below the lower limit, the anions and cations would be farther apart than their equilibrium distance, and triangular coordination would represent a lower energy configuration; above the upper limit, octahedral coordination would result in a lower total energy because more anions would be touching the cation. Calculating the critical radius ratio for each ligancy is quite simple (Problems 2.1 to 2.4), and the results are incorporated into Table 2.2.

The polyhedra which result from connecting the centers of anions surrounding a central cation are called *anion polyhedra,* or, in the more general case of atoms instead of ions, *coordination polyhedra.* The limiting case is for equal-size atoms, as discussed in the previ-

[3] For an explanation of this interesting result, see *Elementary Crystallography,* Chapter 4 (John Wiley, 1956) by M. J. Buerger.

ous section. The coordination polyhedra for HCP and FCC arrangements are shown in Figure 2.4, and should be compared with the atom arrangements shown in Figure 2.2. The coordination polyhedron for both BCC packing and cubic coordination is a cube, but it should be remembered that, by convention, the description, body-centered cubic, is reserved for the case where all atoms are equal size.

A comparison of the calculations of radius ratios at which packing is predicted to change (Table 2.2) and actual radius ratios of a few ionic solids and metals is given in Table 2.3. The data in this latter table were chosen to illustrate some of the exceptions as well as the confirmations of this simple dependency of ligancy on radius ratio. In fact, the agreement is better than one might expect, for very few solids contain bonds that are completely ionic or metallic; some directional covalent bonding is often present. In addition, observed values of ligancy usually are determined from crystals (large aggregates of coordination polyhedra, packed in a three-dimensional, ordered array) in which second-nearest neighbors may influence the ligancy.

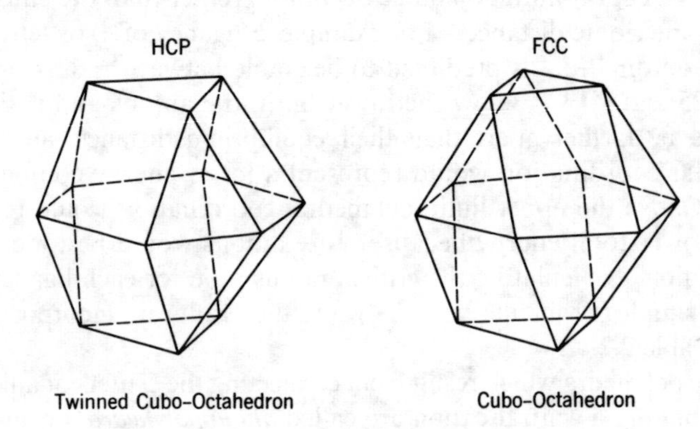

Figure 2.4 Coordination polyhedra for HCP and FCC packing. The cubo-octahedron may be viewed as a cube with its corners truncated by octahedral faces. In the twinned cubo-octahedron, the top half is a mirror image, or twin, of the bottom. The vertices of both polyhedra correspond to the centers of the atoms in Figure 2.2.

Table 2.3 Predicted and Observed Ligancies for Different Ionic Radius Ratios

COMPOUND OR METAL	r_C/r_A	LIGANCY PREDICTED	LIGANCY OBSERVED
B_2O_3	0.14	2	3
BeS	0.17	3	4
BeO	0.23	3 to 4	4
SiO_2	0.29	4	4
LiBr	0.31	4	6
MgO	0.47	6	6
MgF_2	0.48	6	6
TiO_2	0.49	6	6
NaCl	0.53	6	6
CaO	0.71	6	6
KCl	0.73	6 to 8	6
CaF_2	0.73	6 to 8	8
CsCl	0.93	8	8
BCC metals	1.0	8 to 12	8
FCC metals	1.0	8 to 12	12
HCP metals	1.0	8 to 12	12

2.5 THE SIGNIFICANCE OF LOCAL PACKING ARRANGEMENTS

Bonding polyhedra and coordination polyhedra may be visualized as subunits which are stacked together to form the three-dimensional structures of solids. The way in which they stack together determines whether the solid will be noncrystalline (glassy) or crystalline and, if crystalline, what its crystal structure will be. In most cases these polyhedra are simply a descriptive convenience. That is, they are not physical entities but merely more convenient subunits than atoms, and by using them it is possible to consider the local packing arrangements separately from the long-range order of a structure.

A physical as well as a structural significance may be attached to local atomic packing arrangements when the bonding within a group of atoms is stronger than the bonding of this group to the rest of the structure. For instance, a coordination polyhedron behaves as a tightly bound unit when the valence of the central atom is more than half the total valence of the atoms bonded to it. If the valence of the central atom *equals* the total valence of the surrounding atoms, the subunit is really a molecule.

A physical property, such as melting point or softening temperature, which depends on bond strength, will be lower when polyhedral subunits are discrete atomic groups bonded to one another by secondary bonds than when they are bonded together with primary bonds. For example, methane melts at −184°C, ethane melts at −172°C, and higher molecular weight paraffins (each one containing more tetrahedral subunits bonded together with C—C bonds) melt at still higher temperatures. Polyethylene, which is the limiting case of the paraffins, melts at about 125°C. But if adjacent polyethylene chains are bonded together by some C—C bonds produced by irradiation, the polymer may keep its shape and thermal stability up to 300°C. In the extreme case of a three-dimensional network of covalent C—C bonds, diamond is stable to at least 1200°C. As another example, in SiF_4 each silicon is tetrahedrally coordinated by four fluorine atoms, and in SiO_2 each silicon is tetrahedrally coordinated by four oxygen atoms; yet SiF_4 melts at −90°C and SiO_2 melts at 1710°C.

DEFINITIONS

Anion: A negatively charged ion; the result of an electronegative atom's having acquired one or more excess electrons.

Body-Centered Cubic: An atomic packing arrangement in which one atom is in contact with eight atoms identical to it at the corners of an imaginary cube.

Bond Angle: The angle between the two closest directional bonds of an atom.

Bonding Polyhedron: A polyhedron representing the bonding symmetry of a directionally bonded atom; the vertices of the polyhedron indicate the directions of the bonds.

Cation: A positively charged ion; the result of an electropositive atom's having lost one or more of its valence electrons.

Close Packing: An extended three-dimensional packing arrangement in which equal-size atoms are packed as closely together as possible.

Coordination: The packing of atoms around another atom.

Coordination Polyhedron: A polyhedron resulting from connecting the centers of all atoms which are touching a central atom; each vertex represents one of the surrounding atoms; the polyhedron usually is called an anion polyhedron if it represents anions surrounding a cation.

Critical Radius Ratio: The ratio of cation radius to anion radius for the condition where the surrounding anions are touching each other as well as the central cation.

Directional Bonds: Bonds for which the bonding force is greatest in particular directions, for example, all covalent bonds except those involving *s* orbitals, and all permanent dipole bonds.

Double Bond: A covalent bond between two atoms in which each atom uses two half-filled orbitals.

Face-Centered Cubic: An atomic packing arrangement in which twelve atoms surround and touch a central atom of the same species; the resulting coordination polyhedron has cubic symmetry.

Hexagonal Close-Packed: An atomic packing arrangement in which twelve atoms surround and touch a central atom of the same species; the resulting coordination polyhedron has the symmetry of a hexagonal prism.

Ligancy: The number of atoms (or ions) surrounding and touching a central atom (or oppositely charged ion); also called coordination number.

Molecule: A group of atoms bonded to each other with primary bonds and to all other atoms only with secondary bonds.

Nondirectional Bonds: Bonds for which no preferred bond directions exist, for example, metallic bonds, ionic bonds, covalent bonds involving only *s* orbitals, and van der Waals bonds.

Triple Bond: A covalent bond between two atoms in which each atom uses three half-filled orbitals.

BIBLIOGRAPHY

INTRODUCTORY REFERENCES:

W. L. Bragg, *Concerning the Nature of Things,* Dover Publications, New York (1954).

R. T. Sanderson, *Teaching Chemistry with Models,* D. Van Nostrand Co., Princeton (1962), Chapters 1–3.

L. H. Van Vlack, *Elements of Materials Science,* Addison-Wesley Publishing Co., Reading (1959), Chapters 2–3.

SUPPLEMENTARY REFERENCES:

L. V. Azároff, *Introduction to Solids,* McGraw-Hill Book Co., New York (1960), Chapter 4.

R. C. Evans, *Introduction to Crystal Chemistry,* Cambridge University Press (1948).

MORE ADVANCED TEXTS:

L. Pauling, *The Nature of the Chemical Bond,* Cornell University Press, Ithaca (1960).

J. C. Slater, *Quantum Theory of Molecules and Solids,* McGraw-Hill Book Co., New York (1963).

PROBLEMS

2.1 Show that the critical radius ratio for triangular coordination is 0.155.

2.2 Inscribe a tetrahedron in a cube so that the corners of the tetrahedron coincide with four of the corners of the cube. Show that the critical radius ratio for tetrahedral coordination is 0.225.

2.3 Show that the critical radius ratio for octahedral coordination is 0.414.

2.4 Show that the critical radius ratio for cubic coordination is 0.732.

2.5 What factors other than radius ratio might determine the coordination of one atom by another in a crystal?

2.6 Some large anions are easily *polarized* by small, highly charged cations, that is, their electron clouds are distorted so that the charge distribution is no longer symmetrical. How will polarization affect (a) the apparent shape of anions, (b) interionic distances, (c) the general shape of a coordination polyhedron, and (d) the bonding between coordination polyhedra?

2.7 In a BCC crystal the coordination polyhedra are packed face-to-face. Show that in such a crystal there are only two atoms per cube. Calculate the fraction of the total space occupied by spherical atoms in such a crystal.

2.8 Draw a cube with an atom at each corner and one in the center of each face. This is an alternate description of FCC packing, as is shown by the cube in Figure 2.2*b*. If these cubes are packed face-to-face in a crystal, calculate the number of atoms per cube. Calculate the fraction of the total space occupied by spherical atoms in such a crystal. The fraction will be identical with that for ideal HCP structures, since the packing is equally dense in both.

2.9 Calculate the ratio of the volumes of a tetrahedron and an octahedron having the same edge length.

2.10 Calculate the ratio of height to edge-length for a tetrahedron resting on one of its faces. One method of calculating the height is to treat it as a fraction of the body diagonal of a cube in which the tetrahedron is inscribed—as in Problem 2.2.

2.11 Calculate the ratio of cation-anion distance to anion-anion distance in an anion coordination tetrahedron.

2.12 Draw two tetrahedra joined (a) corner-to-corner, (b) edge-to-edge and (c) face-to-face in such a way that the distance between centers is maximized in each case. Compare the distance between centers for each configuration with the single-, double-, and triple-bond lengths betweeen carbon atoms.

2.13 Let the tetrahedra of Problem 2.12 be anion polyhedra. Calculate the ratio of cation-cation distance in each to the anion-anion distance in the tetrahedra.

2.14 The skeleton of a chain of single-bonded carbon atoms may be represented as a chain of tetrahedra joined corner-to-corner. Each tetrahedron may rotate with respect to another tetrahedron about the bond, its edges generating a cone of revolution. Calculate the included half-angle of the cone. If all tetrahedra are randomly oriented with respect to one another, what is the expected configuration of a chain of carbon atoms?

2.15 Compare the rotational freedom of tetrahedra joined corner-to-corner, edge-to-edge and face-to-face. If rotation can occur, specify about how many axes. How must the answer be modified in order that it apply to C—C, C=C and C≡C bonds in a carbon chain?

2.16 Silicones contain long chains of alternating Si and O atoms; their structure may also be represented by tetrahedra joined corner-to-corner. Sketch the skeleton of such a chain. Compare it to the skeleton of a carbon chain.

2.17 Consider the elements Al, Si and S, in Period 3, and Cu, Ge and Se in Period 4 of the Periodic Table (Appendix IA):

	Al	Si	S	Cu	Ge	Se
Density, g/cc	2.70	2.34	2.07	8.96	5.32	4.79
Atomic weight	26.982	28.086	32.064	63.54	72.59	78.96
Metallic character of bond	⟵ increasing			⟵ increasing		
Covalent character of bond	increasing ⟶			increasing ⟶		

Increasing covalent character of the bond produces closer interatomic spacing. As shown above, the covalent character of the bond increases in the same direction as atomic weight. Why does the density decrease with increasing weight?

2.18 If the energy V of a pair of ions of valency Z is taken as zero, the energy of an isolated single ionic molecule is

$$V = \frac{-Z^2e^2}{r} + \frac{B}{r^m}$$

In an ionic crystalline solid, however, each ion has a number of nearest neighbors rather than a single neighbor as in the case of the isolated molecule. Presumably both terms in the expression for total energy will be affected by the presence of more than a single neighbor. Consider a crystal of NaCl ($Z = 1$) in which d_0 is the smallest anion-cation distance. A given ion within the crystal is surrounded by 6 nearest neighbors of unlike sign at a distance d_0, 12 next nearest neighbors of like sign at a distance $\sqrt{2}d_0$, 8 second-nearest neighbors of an unlike sign at a distance $\sqrt{3}d_0$, etc. The total energy of the single ion, considering the coulomb attraction and repulsion of many surrounding ions and the electron cloud repulsion of surrounding nearest neighbors, is then

$$V_0 = \frac{-6Z^2e^2}{d_0} + \frac{12Z^2e^2}{\sqrt{2}d_0} - \frac{8Z^2e^2}{\sqrt{3}d_0} + \frac{6Z^2e^2}{2d_0} - \cdots + \frac{D}{d_0{}^m}$$

$$V_0 = \frac{-MZ^2e^2}{d_0} + \frac{D}{d_0{}^m} = \frac{-MZ^2e^2}{d_0}\left[1 - \frac{1}{m}\right]$$

The constant M is the *Madelung constant;* it depends on the crystal geometry and for the NaCl structure has the value

$$M = \frac{6}{\sqrt{1}} - \frac{12}{\sqrt{2}} + \frac{8}{\sqrt{3}} - \frac{6}{\sqrt{4}} + \frac{24}{\sqrt{5}} - \cdots$$

This series converges at an extremely low rate, and several simple, approximate techniques have been devised for summing it. One such technique is as follows: consider a cube, whose edges are rows of ions, surrounding a central reference ion; compute the coulomb term for each ion or part ion lying within the cube. An ion lying in a cube face is considered a half ion, one lying on a cube edge is considered a quarter ion, and one lying at a cube vertex is considered a one-eighth ion. Compute the Madelung constant for NaCl (a cube with edge $4d_0$ will give a result within one quarter of one per cent of the precise value). The first few terms are

$$M = \frac{6}{\sqrt{1}} - \frac{12}{\sqrt{2}} + \frac{8}{\sqrt{3}} - \frac{6/2}{\sqrt{4}} + \frac{24/2}{\sqrt{5}} - \cdots$$

2.19 Explain why SiO_2 melts at a much higher temperature then SiF_4.

CHAPTER THREE

Crystal Structure

Crystal structures are regular, three-dimensional patterns of atoms in space. The regularity with which atoms usually are packed in solids arises from geometrical conditions which are imposed by directional bonding and close packing. Crystal structures observed in solids are described in terms of an idealized geometric concept called a space lattice and may be rationalized in terms of the way coordination polyhedra pack together to minimize the energy of the solid.

3.1 INTRODUCTION

Explaining why a given group of atoms crystallizes in one structure rather than another is complicated, in all but the simplest cases, by our ignorance of the exact details of atomic bonding. In fact, if we were to try to predict crystal structures from what we know presently about the way specific atoms bond together under all circumstances, we would probably be wrong almost as often as right. However, we can state some simple rules describing what factors we expect to be important in determining the packing of coordination polyhedra (or atoms) and then rationalize observed crystal structures in terms of these rules. This procedure is slightly artificial, but it provides a framework within which we can consider how individual crystal structures are related to one another, geometrically as well as chemically.

Ideally, the most stable arrangement of coordination polyhedra in a crystal will be that which minimizes the energy per unit volume or, in other words, the one that

1. preserves electrical neutrality

2. satisfies the directionality and discreteness of all covalent bonds
3. minimizes strong ion-ion repulsion
4. packs the atoms as closely as possible, consistent with (1), (2), and (3).

Before discussing the actual three-dimensional patterns made by atoms in a crystal structure, it is useful to consider briefly what patterns are possible for identical *points* in space. Such patterns of points are called *space lattices,* and every crystal structure is based on one of the possible space lattices.

3.2 THE BRAVAIS SPACE LATTICES

A space lattice is an infinite, three-dimensional array of points in which every point has surroundings identical with that of every other point. These points with identical surroundings are called *lattice points.* Lattice points can be arranged in only 14 different arrays, called Bravais lattices; therefore the atoms in any crystal structure must be in positions designated by 1 of the 14. It is important to note that more than one atom can be associated with each lattice point; but for every atom or group of atoms at one lattice point, there must be an identical atom or group of atoms with the same orientation at every other lattice point to satisfy the definition of a space lattice.

Since the structure of a perfect crystal is a regular pattern of atoms, distributed on a space lattice, the atomic arrangements can be described completely by specifying atom positions in some *repeating* unit of the space lattice. Such a repeating unit of the space lattice is called a *unit cell.* If atom positions are specified within the unit cell, we shall call it a *crystal-structure unit cell.* The edges of a unit cell must be *lattice translations* (vectors which connect any two lattice points), and identical unit cells of a particular space lattice will fill space and generate the space lattice when they are packed face-to-face. One space lattice can have a number of different unit cells that satisfy the criteria above, but, conventionally, unit cells are chosen which have a simple geometry and contain only a few lattice points. Examples are the unit cells of the 14 Bravais lattices shown in Figure 3.1.

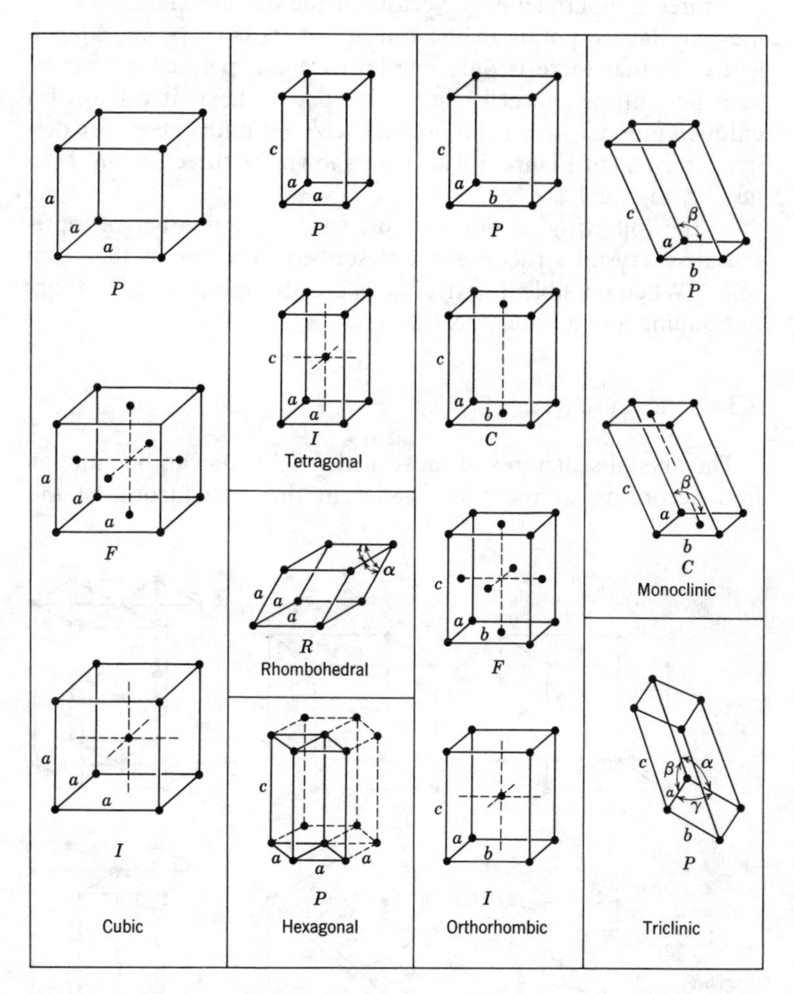

Figure 3.1 Conventional unit cells of the 14 Bravais space lattices. The capital letters refer to the type of cell—*P:* primitive cell; *C:* cell with a lattice point in the center of two parallel faces; *F:* cell with a lattice point in the center of each face; *I:* cell with a lattice point in the center of the interior; *R:* rhombohedral primitive cell. All points indicated are lattice points. There is no general agreement on the unit cell to use for the hexagonal Bravais lattice; some prefer the *P* cell shown with solid lines, and others prefer the *C* cell shown in dashed lines.

If three nonparallel edge vectors of the cell are chosen so that the only lattice points in the cell are at its corners (or, equivalently, so that there is only one lattice point per cell, as can be seen by shifting the cell slightly in space), the unit cell is of a unique kind and is called a *primitive cell.* Primitive cells are designated by P in Figure 3.1 and are shown for three *crystal structures* in Figure 3.2.

In the following sections of this chapter, a number of representative crystal structures are described in terms of their unit cells. When possible, the structures are also rationalized in terms of bonding and atomic packing.

3.3 THE ELEMENTS

The crystal structures of most of the elements can be rationalized according to the rules stated in the introduction of this

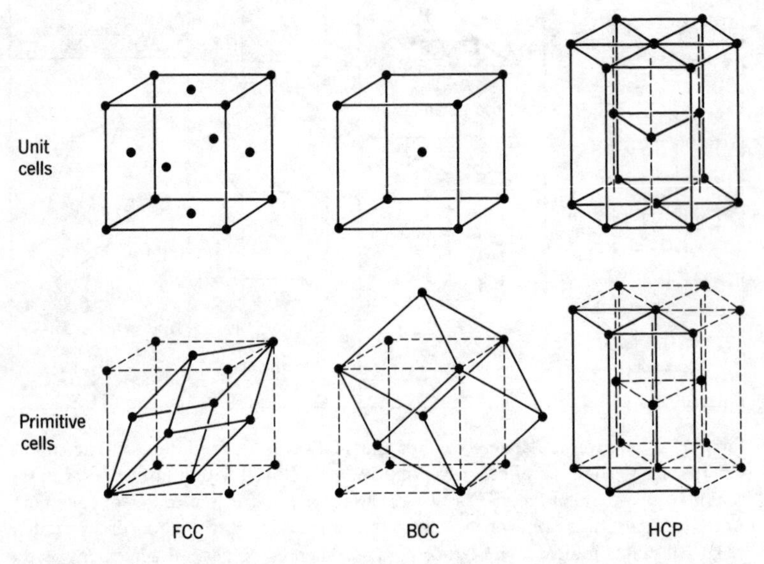

Figure 3.2 The unit cells and the primitive cells of FCC, BCC, and HCP crystal structures. Points represent *atom positions* but are lattice points also in the FCC and BCC structures. In the HCP cell, the atom positions *within* the cell are not lattice points.

chapter. Since we are concerned here only with the elements, electrical neutrality already exists, and there is no ion-ion repulsion. Thus the two requirements to be considered are the satisfaction of covalent bonds and the close packing of atoms.

First, let us consider the true metals and the noble elements. The conventional unit cells and the associated primitive cells of the three most common metallic and noble-element crystal structures are shown in Figure 3.2.[1] For these elements the primary condition to be satisfied is that of packing the atoms as closely together as possible, since the bonding is not directional. One way to picture the close packing of atoms is in terms of their coordination polyhedra. Atoms are packed as closely as they can be when their coordination polyhedra are stacked together as closely as possible. In fact, the FCC, HCP, and BCC structures can be derived by stacking their respective coordination polyhedra together so they share the maximum number of faces. The BCC arrangement is simply an array of stacked cubes. The FCC and HCP arrangements are more difficult to visualize in this way and are seen more easily as a result of atomic packing.

Since FCC and HCP arrangements of atoms are the densest possible arrangements for spherical atoms, another way to view them is in terms of *close-packed planes* of atoms stacked to minimize the volume they occupy. A close-packed plane is a two-dimensional array of atoms packed together as closely as possible. Two close-packed planes stacked to minimize their total volume are shown in Figure 3.3. If a third plane is stacked on these two in such a way that its atoms are directly above those in the first plane—compare Figure 2.2*a*—and this sequence is continued, the structure is HCP. The *stacking sequence* of close-packed planes in an HCP crystal structure is described as ABABA.... Alternatively, if the third plane is stacked so that its atoms are directly above the "holes" between the atoms in both first and second planes—compare Figure 2.2*b*—and this sequence is continued, the structure is FCC. The stacking sequence of close-packed planes in an FCC crystal structure is described as ABCABCABCA....

[1] Note that in the HCP crystal structure, two atoms are associated with each lattice point of the hexagonal Bravais lattice, and therefore the primitive cell of the crystal structure contains two atoms but only one lattice point.

Figure 3.3 One close-packed plane of atoms (shown as semitransparent spheres) above another plane (shaded spheres). The planes are closest to each other when the atoms in one rest over the holes between the atoms of the other.

An equivalent view of the HCP, FCC, and BCC crystal structures is shown in Figure 3.4 in which the atoms are represented by spheres, and each figure represents a number of unit cells of one crystal structure packed together—compare Figure 3.2. Note that a corner has been "sliced" off the cube representing FCC packing to expose one of the close-packed planes and show its orientation. The exposed plane is outlined with a triangle. In the HCP models, the close-packed planes are parallel to the base. As noted in the previous chapter, no metal has an ideal HCP structure. The close-packed planes are always a little farther apart or a little closer together than close-packed spheres would be (see Problem 3.13).

Next, let us consider some of the elements with more than three valence electrons. These elements have predominantly covalent bonds and, therefore, crystallize in structures which satisfy these bonds. When all the unfilled orbitals are used for covalent bonding, each atom surrounds itself with $(8 - N)$ neighboring atoms, where N is the number of outer shell, or valence, electrons. As a result, an atom such as Cl, Br, or I, with an outer shell containing seven electrons, is bonded primarily to only one other like atom, forming a diatomic molecule. These molecules are bonded to the other molecules in a crystal (Figure 3.5) by weak secondary bonds. The atoms of group VI (S, Se, Te) have outer shells with six electrons and form chain or ring molecules so that each atom has only two close neighbors in positions which satisfy the bond

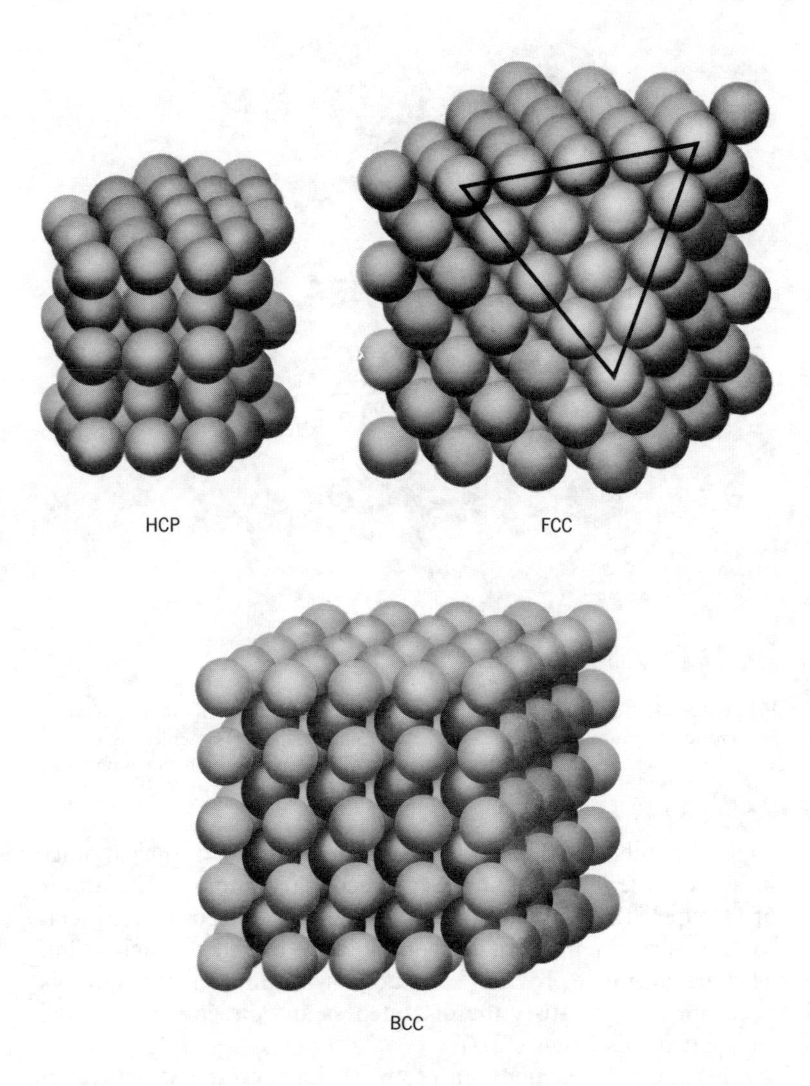

Figure 3.4 The HCP, FCC, and BCC crystal structures. Atoms are represented by spheres. In the FCC structure, one corner has been "sliced" off to expose a close-packed plane and show its orientation within the cube. The exposed close-packed plane is outlined with a heavy triangle.

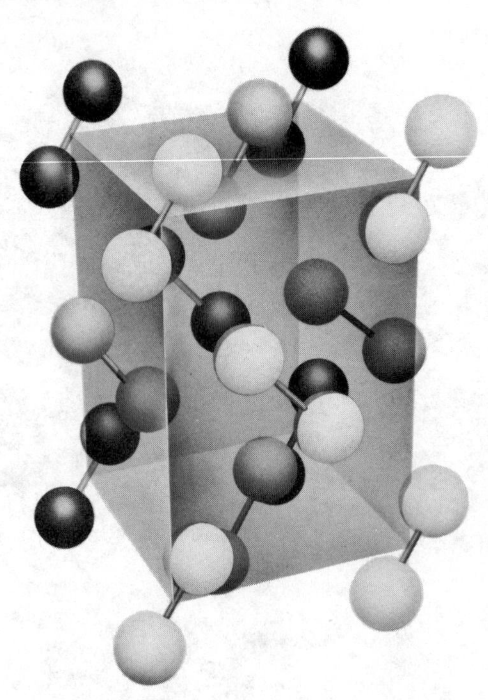

Figure 3.5 The crystal structure of iodine, showing the packing of I_2 molecules referred to an orthorhombic unit cell. Atom positions are represented by spheres and directional bonds by rods between spheres. The atomic coordination is one-fold.

angle (Chapter 2). The molecules are then bonded to each other in a crystal (Figure 3.6) by weak secondary bonds. The atoms of Group V (P, As, Sb, Bi) have five valence electrons and tend to crystallize in puckered sheets, loosely bound to each other. This arrangement provides each atom with three close neighbors in positions that satisfy the discreteness and directionality of the covalent bonds (Figure 3.7).

Only the light elements in group IV have crystal structures in which *all* the bonds that hold the crystal together are covalent. The bonds are the result of overlapping sp^3 hybrid orbitals and the structure that results is the diamond cubic (DC) crystal structure of diamond, Si, Ge, and grey Sn, shown in Figure 3.8. Some of the group-IV elements can also have structures in which all the bonds that hold the crystal together are not covalent. Graphite is one crystalline form of carbon; in it the atoms are covalently bonded in planar hexagonal arrays with weak secondary bonds between the planes (Figure 3.9). The primary covalent bonds in this case are thought to be between sp^2 orbitals; the weaker bonds between the planes are much more metallic in character with

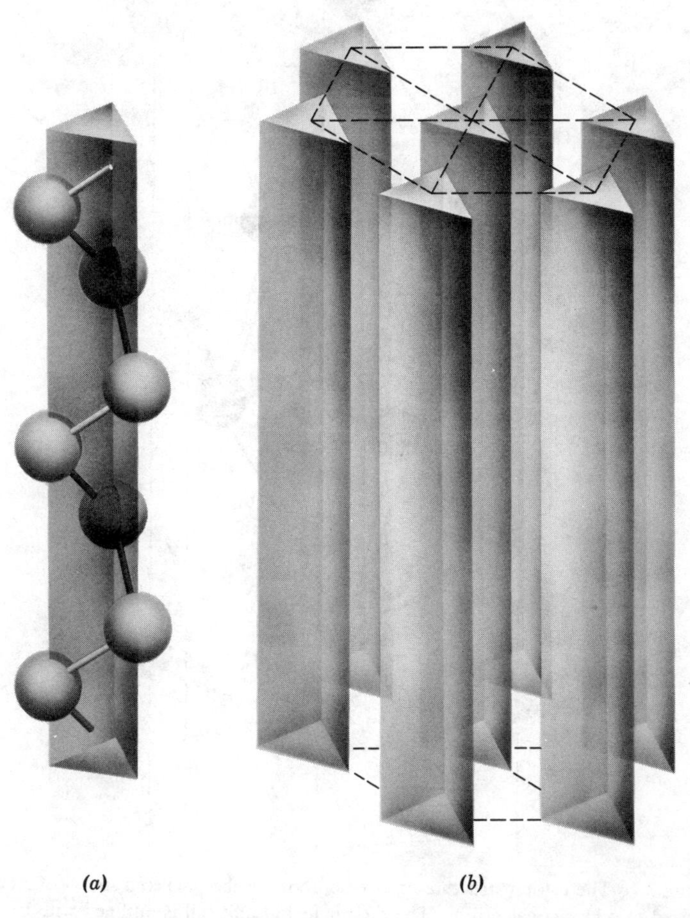

Figure 3.6 The crystal structure of tellurium, showing: (*a*) a spiral chain of atoms, referred to a triangular prism, and (*b*) the arrangement of these prisms on a hexagonal lattice. Atom positions are represented by spheres and directional bonds by rods between spheres. The atomic coordination is two-fold.

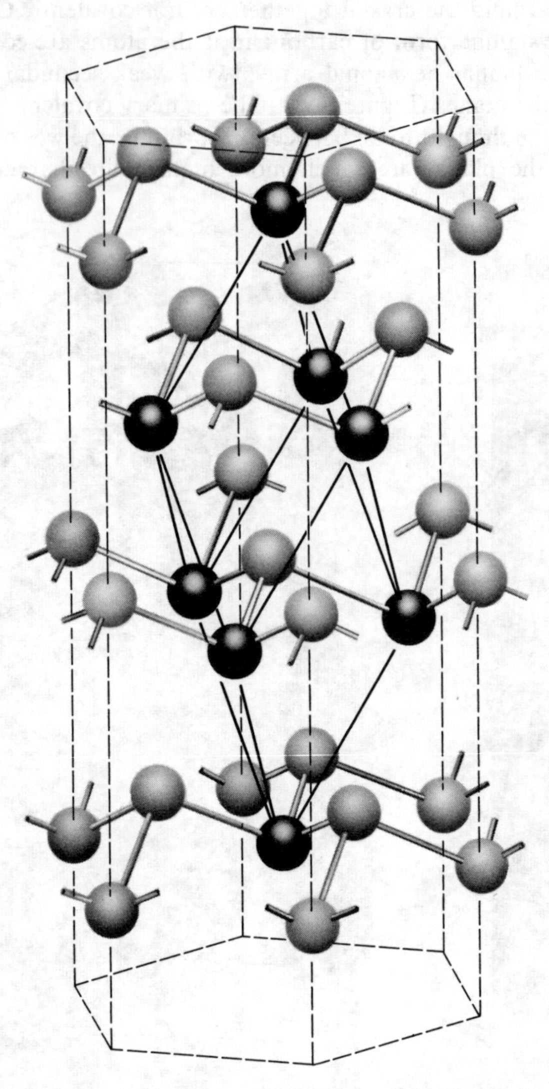

Figure 3.7 The crystal structure of arsenic, showing the puckered sheets of atoms referred to a hexagonal prism. The rhombohedral unit cell is outlined with heavy lines and darker atoms. Atom positions are represented by spheres and directional bonds by rods between spheres. The atomic coordination is three-fold.

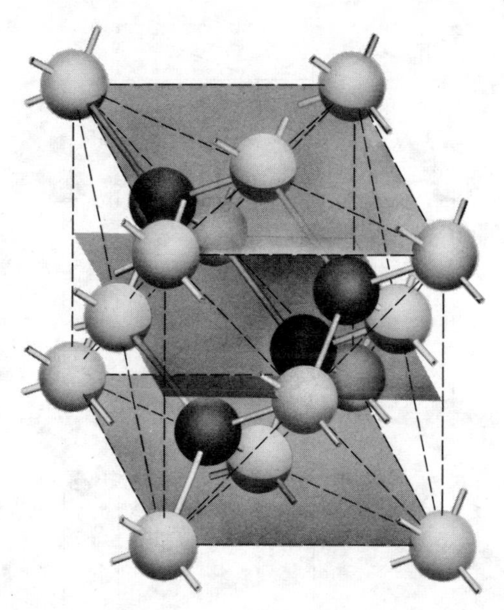

Figure 3.8 The diamond cubic crystal structure. Atom positions are represented by spheres and directional bonds by rods between the spheres. The positions of the light spheres show that the diamond cubic structure is a FCC lattice with two atoms per lattice point. In this figure the two types of atoms with different surroundings are represented by light spheres (on FCC lattice points in the unit cell) and by dark spheres (tetrahedrally coordinated by the light spheres, and inside the cell). The diagonal planes across which the carbon atoms appear to be widely spaced are the cleavage planes of diamond.

the result that graphite conducts electricity and heat much more easily parallel to the planes than perpendicular to them. In white (metallic) tin, the bonding is also mixed between metallic and covalent, and the crystal structure is body-centered tetragonal with two atoms per lattice point. Lead has a typically metallic crystal structure (FCC).

3.4 IONIC CRYSTALS

Although very few crystals are completely ionic, many have a significant degree of ionic bonding and can be classified, broadly, as *ionic crystals*. Examples are NaCl, MgO, SiO_2, and LiF. In

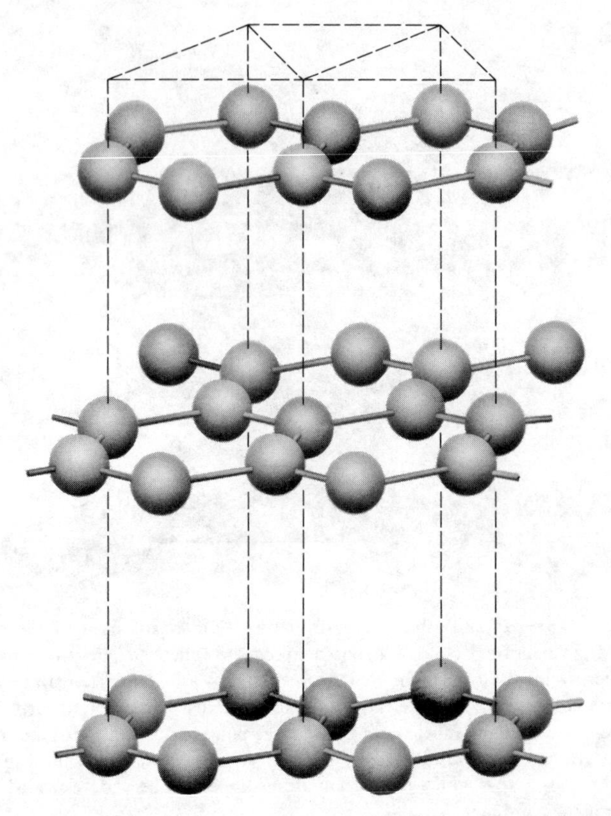

Figure 3.9 The crystal structure of graphite, showing the hexagonal sheets of carbon atoms. Atom positions are represented by spheres and directional bonds by rods between spheres. The atomic coordination is three-fold.

purely ionic crystals, the anion polyhedra are packed together so as to maintain electrical neutrality and minimize the bonding energy per unit volume without inducing any strong repulsion between like-charged ions. The repulsion is largest between cations because their charge is concentrated in a smaller volume, and therefore the polyhedra must be stacked in such a way that the cations at the centers are far enough apart so that the repulsion is negligible. Small coordination polyhedra (triangles and tetrahedra) around highly charged cations usually share only corners with each other, thus maximizing the distance between cations. They are less likely to share edges and are never found stacked face-to-face. As the ligancy increases and the cation charge decreases, the anion polyhedra can be packed more closely together—that is, they may share edges or even faces—without bringing the cations too close to each other. The arrangements of coordination polyhedra in a number of representative crystal structures described below illustrate these points.

Triangular coordination is rarely observed in ionic solids because it requires such a large difference in the sizes of the ions. One example is boron oxide, B_2O_3; however, this compound is seldom found in crystalline form. The polyhedra are triangles and undoubtedly share only corners; but the closeness of the boron ions to one another and the difficulty of building a stable three-dimensional structure from triangles joined at their corners probably explain why the compound is usually noncrystalline (see Chapter 5).

Crystal structures containing either tetrahedrally or octahedrally coordinated cations which are not highly charged can be discussed within the same geometric framework for the following reason. If the cations are not highly charged, the anion polyhedra will be stacked as closely together as possible to minimize the energy per unit volume; and a close-packed, repetitive array of tetrahedra contains *octahedral* spaces between the tetrahedra. Similarly, a close-packed, repetitive array of octahedra contains *tetrahedral* spaces. This is another way of stating that neither regular tetrahedra nor regular octahedra will, by themselves, fill space completely, but that a combination of the two will do so. Therefore we shall consider the two together. Crystallographically, this combination of tetrahedra and octahedra results in placing anions on the atomic sites of either a FCC or a HCP structure, with the cations either in tetrahedrally or octahedrally coordinated sites between them. Although this is not obvious at first, Figure 3.10 shows one site of each type between two close-packed planes of spheres. If the centers of all the neighboring spheres in the figure are connected by straight lines, the result will be a layer of tetrahedra and octahedra packed face-to-face. Adding a third plane of spheres corresponds to adding another layer of tetrahedra and octahedra. If tetrahedra are stacked on tetrahedra and octahedra on octahedra (if similar polyhedra have

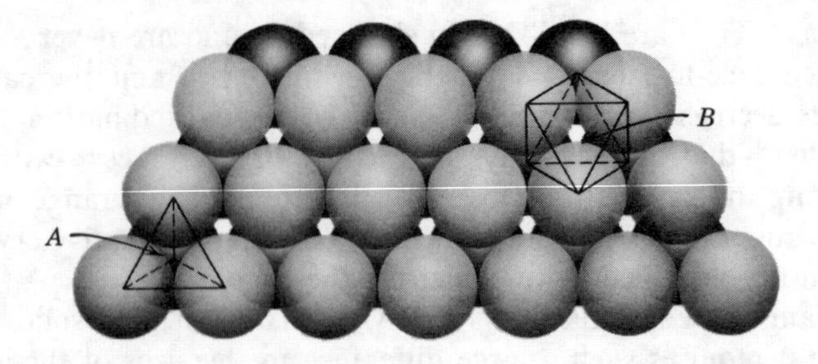

Figure 3.10 One close-packed plane of spheres stacked on top of another one. A tetrahedral interstitial site (*A*) and an octahedral interstitial site (*B*) are indicated. All of the octahedral sites have the same orientation, but some of the tetrahedral sites are such that the tetrahedron is upright; some are such that the tetrahedron is inverted, as in (*A*).

faces in common), the stacking of the spheres representing anion sites is HCP. If tetrahedra are stacked on octahedra and vice versa (if similar faces have *only* edges in common), the stacking of the spheres is FCC.

It is important to realize that, in this representation of crystal-structure geometry by tetrahedra and octahedra, the vertices of the polyhedra represent atomic *sites* in a close-packed crystal structure. The differences between representing FCC- and HCP-packing by an arrangement of spheres and by an arrangement of tetrahedra and octahedra are shown in Figures 3.11 and 3.12.

In an ionic crystal, the vertices represent anion sites,[2] and the centers of tetrahedra and octahedra represent cation sites. In a true ionic crystal, whether tetrahedral or octahedral sites are occupied is determined by the ratio of the ion radii: tetrahedral coordination is stable for $0.225 < r_C/r_A < 0.414$, and octahedral coordination is stable for $0.414 < r_C/r_A < 0.732$. The fraction of available cation sites that will be occupied depends only on the chemical formula, since there are fixed numbers of each type of site. The geometry of packing in these arrangements is such that there are equal numbers of close-packed anion sites, octahedral

[2] It should be emphasized that, in such a binary compound crystal, although the anions are on close-packed *sites* (FCC or HCP) they will not actually be in contact with each other if the cation size is larger than that given by the *critical* value of the radius ratio for the type of coordination in question.

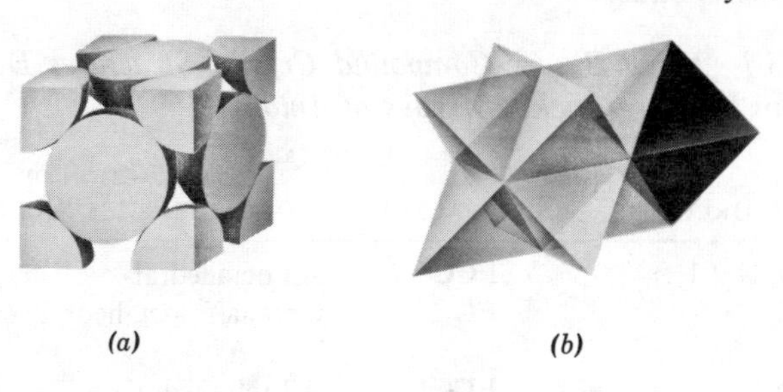

Figure 3.11 (*a*) A cubic unit cell cut out of a FCC array of spheres. Only the portion of each sphere actually in the unit cell is indicated. (*b*) The unit cell of (*a*), represented by the geometric figures which result when the centers of neighboring spheres are connected by straight lines. The figures are both tetrahedra (shaded light) and octahedra (shaded dark). Other octahedra (not shown) fit along each edge of the cubic unit cell. Note that this representation focuses attention on the coordination of the interstitial sites in a close-packed array.

cation sites, "upright" tetrahedral cation sites, and "inverted" tetrahedral cation sites. Some common crystal structures of binary compounds that fit into this geometric pattern are described in Table 3.1. In those compounds that form true ionic crystals, the arrangement of cations that occupy only a fraction of one type of site is that which maximizes the average cation–cation distance. For those crystals in which the bonding is not

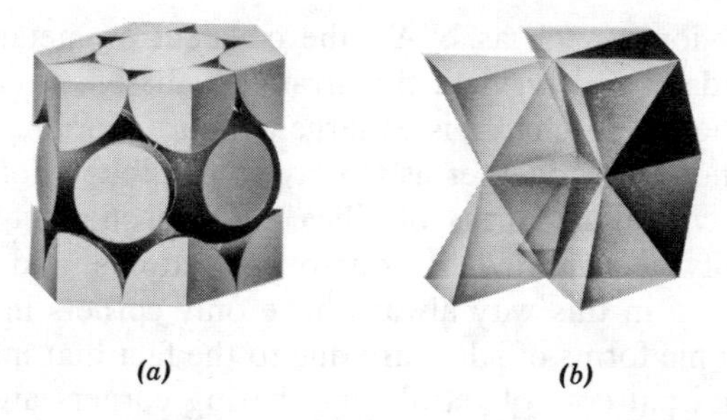

Figure 3.12 (*a*) A hexagonal unit cell cut out of a HCP array of spheres. Only the portion of each sphere actually in the unit cell is indicated. (*b*) The unit cell of (*a*), represented by the geometric figures which result when the centers of neighboring spheres are connected by straight lines. The figures are both tetrahedra (shaded light) and octahedra (shaded dark). Other octahedra and tetrahedra (not shown) fit along the edges of the unit cell. Compare the stacking of polyhedra in this figure with the stacking shown in Figure 3.11.

Table 3.1 Some Binary-Compound Crystal Structures Described in Terms of Close-Packed Arrays of Anions

CRYSTAL STRUCTURE	ANION PACKING	FRACTION OF CATION SITES OCCUPIED
MgO, NaCl	FCC	All octahedral
$CdCl_2$	FCC	One-half octahedral (alternate layers)
K_2O[a]	FCC	All tetrahedral
ZnS (cubic)	FCC	All upright tetrahedral
$Zn(CN)_2$[b]	FCC	One-quarter upright tetrahedral plus one-quarter inverted tetrahedral
NiAs	HCP	All octahedral
Al_2O_3	HCP	Two-thirds octahedral
CdI_2	HCP	One-half octahedral (alternate layers)
ZnS (hexagonal)	HCP	All upright tetrahedral

[a] The K_2O structure is commonly called "antifluorite" because in fluorite (CaF_2) the geometry is the same but the cations, rather than the anions, are on the close-packed sites. Note that in the fluorite arrangement the coordination of anions around cations is cubic, as predicted. See Table 2.3.
[b] The $Zn(CN)_2$ structure is commonly called "anticuprite" because in cuprite (Cu_2O) the geometry is the same but the cations are on the close-packed sites.

completely ionic, such as NiAs, the covalent or metallic nature of the bonds may determine the arrangements of the cations.

When the cation charge is as large as four, coordination tetrahedra do not pack together as closely as possible. For example, in all the crystalline forms of silica, SiO_2, each silicon atom is tetrahedrally coordinated by four oxygen atoms, and the tetrahedra formed in this way always have only corners in common. The allotropic forms of silica are due to the fact that many three-dimensional patterns of tetrahedra sharing corners are possible. Unit cells of two of these arrangements, the crystal structures of β-quartz and cristobalite, are shown in Figure 3.13. In β-quartz the tetrahedra are joined corner-to-corner in such a way that they spiral around a hexagonal prism. In cristobalite, the silicon atoms are arranged on a diamond cubic lattice with oxygen atoms between each pair of nearest silicon atoms.

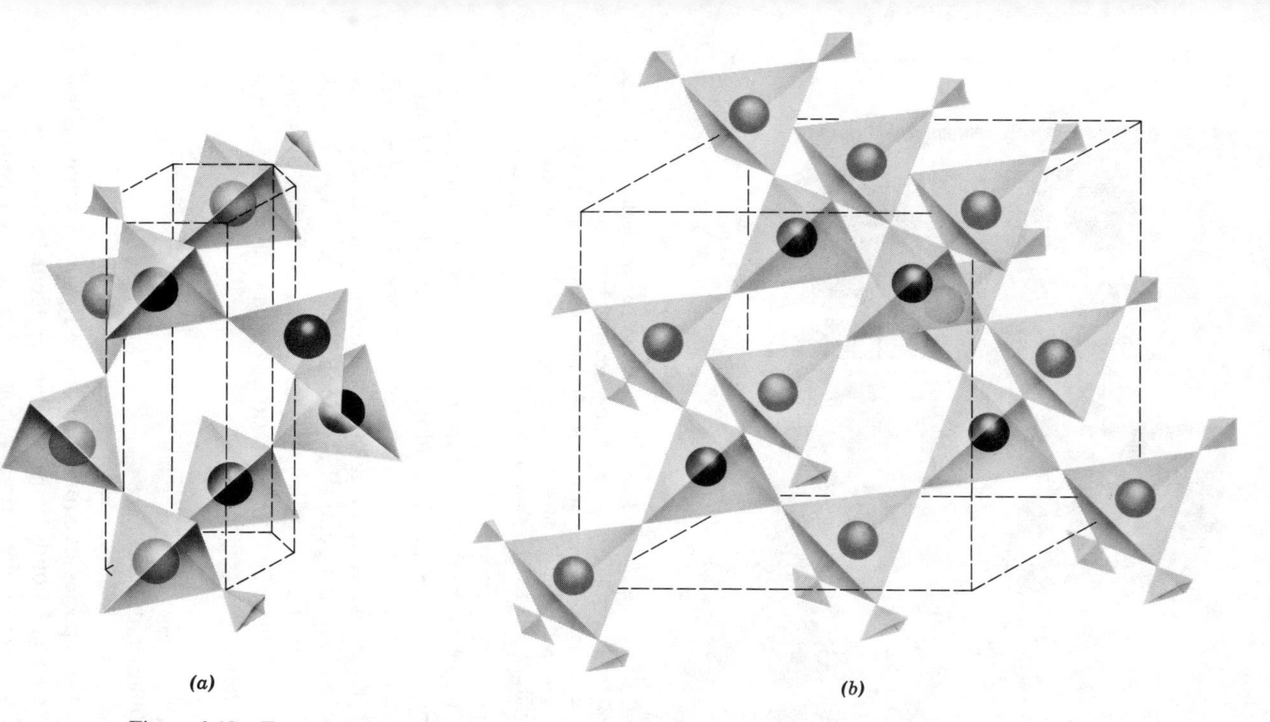

Figure 3.13 Two of the crystal structures of SiO_2, showing the arrangements of the Si—O tetrahedra. (*a*) The crystal structure of β-quartz in which two spiral chains wind around a hexagonal prism. The silicon atoms in one chain are shaded darker than in the other chain to distinguish between the two. (*b*) The crystal structure of cristobalite in which the tetrahedra are arranged similarly to the carbon atoms in diamond. The silicon atoms that are on FCC lattice points in the unit cell are shaded somewhat lighter than those lying within the cell. Compare with Figure 3.8.

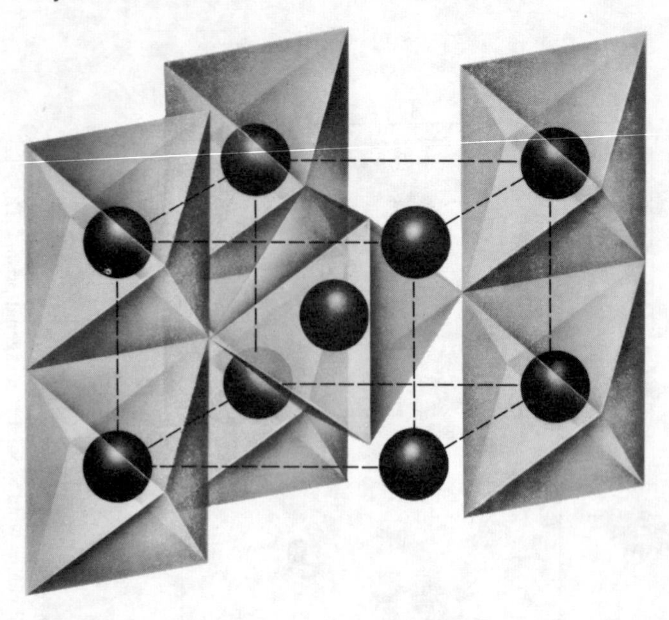

Figure 3.14 The arrangement of the Ti—O coordination octahedra in the rutile structure of TiO_2. For clarity, the octahedra around two titanium atoms on one front edge of the unit cell are not shown. Their orientations are exactly the same as those on the other three edges of the unit cell. Note that the cell is tetragonal and that the octahedra share corners and edges, but not faces.

An example of octahedral coordination when the cation charge is as large as four is titanium dioxide (TiO_2). The Ti—O octahedra are large enough that they can be packed closely enough to share edges without bringing the cations too close to each other. However, the octahedra do not share faces. One crystallographic form of TiO_2, called rutile, is shown in Figure 3.14, and it can be seen that the octahedra do not even share all edges, just some of them.

3.5 MOLECULAR CRYSTALS

When two species of atoms are bonded primarily with either ionic or covalent bonds, it is possible for them to form discrete molecules. When the primary bonds are satisfied completely within a subunit, the subunits must then be held together by a type of bond different from the primary bond. In such *molecular crystals,* subunits are held together with weak secondary bonds.

When the primary bonding is mostly ionic, discrete molecules are formed if the cation charge equals the product of the anion charge and the ligancy. An example of this type of molecular crystal is SiF_4, in which the tetrahedra (of F around Si) are arranged on a BCC lattice. Another example is the mineral brucite, Mg $(OH)_2$, in which the Mg^{+2} ions are octahedrally surrounded by $(OH)^-$ ions. The coordination octahedra lie in a sheet such that each $(OH)^-$ ion is shared by three Mg ions. In a sheet containing N Mg^{+2} ions, there are $6N/3 = 2N$ $(OH)^-$ ions. Such a sheet is, therefore, a molecule and is loosely bonded to other sheets by van der Waals bonds.

In ice, the primary bonds are covalent, and the secondary bonds are weak ionic (hydrogen) bonds. Hydrogen can form only one covalent bond because its only stable bonding orbital is the 1*s* orbital. Thus a molecule of water is an oxygen atom with its two covalent bonds satisfied by two hydrogen atoms. The angle between hydrogen atoms normally is about 105° (see Chapter 2). In the solid, crystalline state the hydrogen atoms bond the molecules together by means of the ionic, or dipole hydrogen bond. The actual arrangement of water molecules in an ice crystal is shown schematically in Figure 3.15 and shows the origin of the hexagonal symmetry of snowflakes. Since the H—O—H bond angle is close to the tetrahedral angle (109°28′), the structure may be visualized as a ZnS (hexagonal) structure (see Section 3.4) in which both Zn and S are replaced by O, and the hydrogen ions (protons) are located on lines joining nearest oxygens. Each hydrogen atom is always closer to one of the oxygen atoms than to the other, preserving the molecular nature of the crystal. It is possible, alternatively, to view the oxygens as being on the atom sites of a hexagonal analogue of the diamond cubic structure.

The largest class of molecular crystals is that in which covalently bonded molecules have weak, van der Waals, intermolecular bonds. When the molecules are approximately spherical, either because of the shape of the coordination polyhedra or because of free molecular rotation, the crystal is usually a close-packed array of these molecules held together with nondirectional

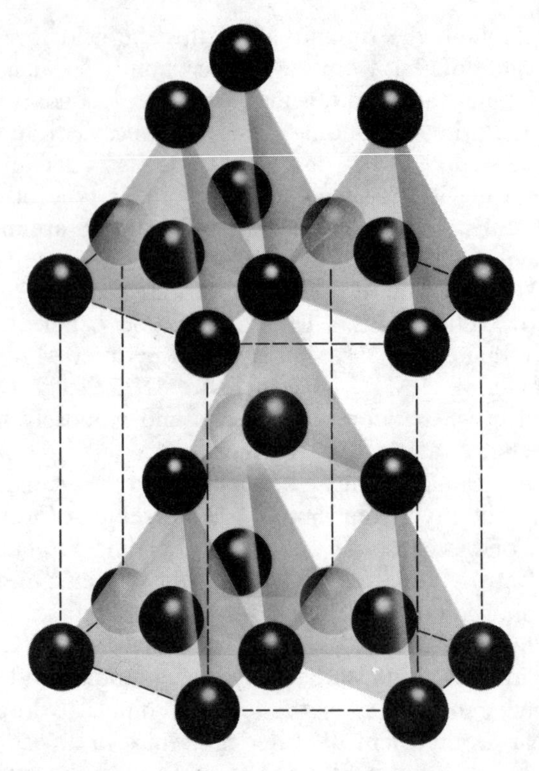

Figure 3.15 The tetrahedral arrangement of oxygen atoms (ions), represented by spheres, in the crystal structure of ice. Hydrogen ions are not shown, but are located on lines joining nearest oxygen atoms in such a way that each oxygen atom has two hydrogens closest to it.

van der Waals bonds. This occurs, for example, in crystals of CH_4 and NH_3 at low temperatures where the molecules form a FCC array.

Of more engineering importance are the crystals of long chain molecules. When the molecules are long chains with reasonably symmetrical cross sections, they usually crystallize on an orthorhombic or a monoclinic lattice. In the monoclinic arrangement, all chains are parallel to each other but inclined to the base of the unit cell. The orthorhombic arrangement results when the axes of the chains are parallel but their planes are inclined in two different directions. The packing of the polymer, polyethylene, is an example of orthorhombic packing and is shown in Figure 3.16. Although polymers like polyethylene often have noncrystalline regions because the chains are so long that they become tangled easily, small single crystals have been grown from solution. In such crystals, the carbon chain bends back on itself every few hundred Angstroms, but the *local* packing is that described in Figure 3.16.

Most polymers which have more than two kinds of atoms have

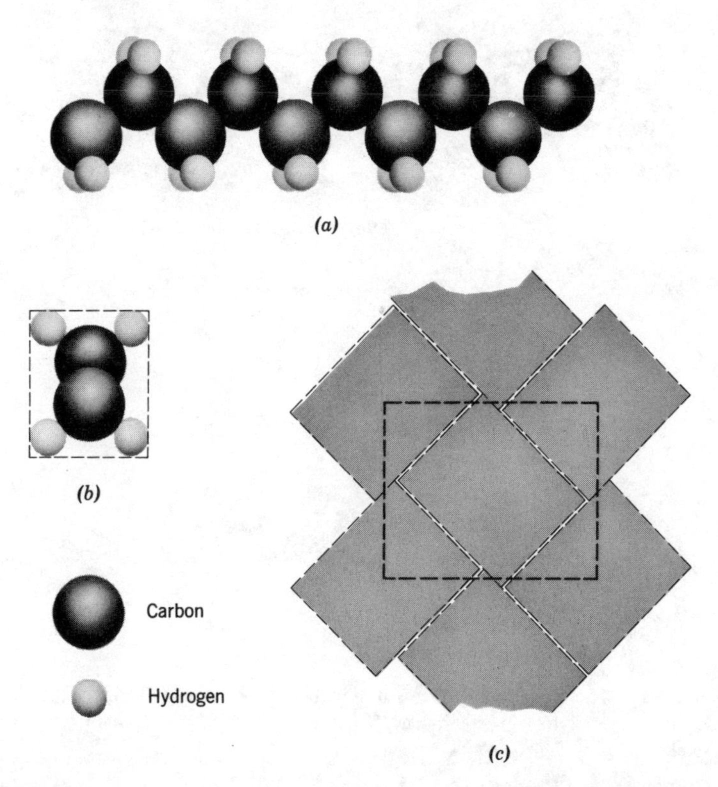

Figure 3.16 The molecular and crystal structure of polyethylene. (*a*) The planar zig-zag arrangement of the carbon chain backbone of the structure can be seen. All the carbon bonds are directed tetrahedrally. (*b*) An end view of a polyethylene chain, showing its approximately rectangular shape. (*c*) The arrangement of polyethylene chains in an orthorhombic crystal of the polymer. This is an end view, the third dimension being determined by the repeat distance along the chain. The heavy dashed line outlines the end face of the unit cell.

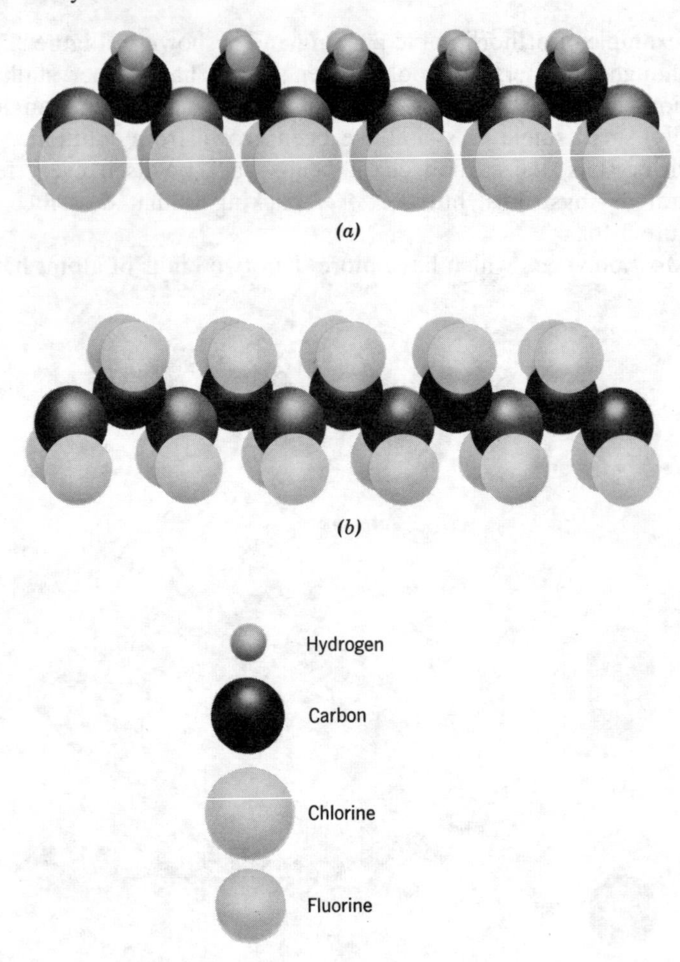

Figure 3.17 The atom arrangements in polymer chains of (*a*) polyvinylidene chloride and (*b*) polytetrafluoroethylene. The regularity of each chain promotes crystallization.

such irregular chains that they do not crystallize. However, there are some, such as polyvinylidene chloride and polytetrafluoroethylene (Teflon) which have such symmetrical cross sections (Figure 3.17) that they crystallize easily. It is also possible for a polymer to be chemically complex and still have such a sym-

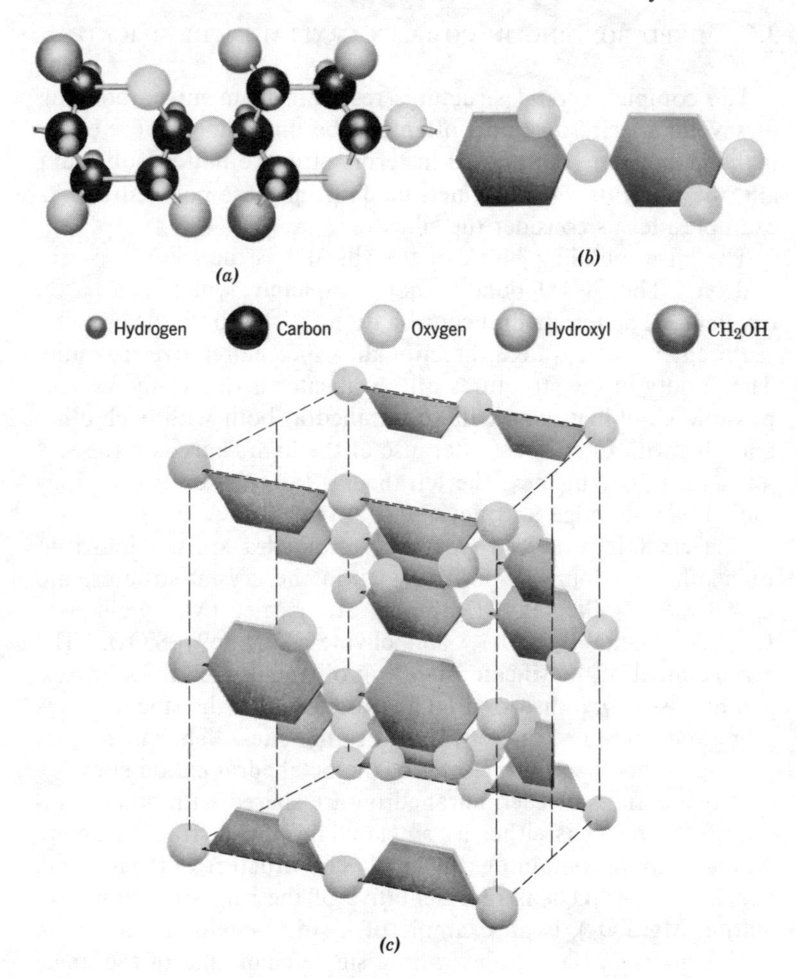

Figure 3.18 (*a*) The atomic arrangement within a chain of cellulose. (*b*) A more schematic representation of the cellulose chain. (*c*) A schematic representation of the arrangement of cellulose chains in a crystal.

metrical structure that it crystallizes. An example is cellulose (Figure 3.18) found in wood and in natural plant fibres like cotton. The way in which the oxygen bonds the hydrocarbon rings together keeps the cellulose chains straight and promotes crystallization.

3.6 MIXED BONDING IN COMPLEX CRYSTALS: THE SILICATES

The complex crystal structures resulting from mixed bonding of several species of atoms often can be understood more easily if the crystals are considered in terms of coordination polyhedra and the subunits, or molecules, made from these polyhedra. As examples, let us consider the silicates.

The basic building block of the silicates is the $(SiO_4)^{-4}$ tetrahedron. The Si—O bond is not completely ionic; it is partly covalent. The tetrahedral coordination satisfies both the bonding requirements of covalent directionality and the relative size ratio. The variety in the structures of the silicates is due to the various possible combinations of these tetrahedra, both with each other and also with other ions. Because of the high charge on the Si^{+4} ion and its low ligancy, the tetrahedral $(SiO_4)^{-4}$ units are rarely joined edge-to-edge and never face-to-face.

The crystals produced when the tetrahedra are not joined to each other but only to positive ions in the crystal structure are called *island* silicates. Examples are garnet $(Mg, Fe^{II}, Mn, Ca)_3(Cr, Al, Fe^{III})_2(SiO_4)_3$[3] and olivine $(Mg, Fe^{II})_2(SiO_4)$. The arrangement of the silicate islands in olivine is such that the oxygen atoms form a distorted HCP structure with the silicon atoms filling one-eighth of the tetrahedral cation sites. The magnesium or iron atoms then fill one-half of the octahedral cation sites.

If two corners of each tetrahedron are shared with other tetrahedra, the result is either a "single chain" or a "ring" structure with a formula containing $(SiO_3)^{-2}$. The structure of the mineral, beryl, $Be_3Al_2(SiO_3)_6$ is representative of the ring structure. Enstatite, $Mg(SiO_3)$, is an example of a single-chain silicate. The arrangement of the tetrahedra in a single chain and of the single chains in a crystal are shown in Figure 3.19. Cleavage planes (planes across which the crystal breaks apart easily) in single-chain silicates intersect at specific angles (87° and 93°) because it is easier to break the ionic bonds between silicate chains and positive ions than it is to break Si—O bonds.

[3] Elements separated by commas are found substituted for each other in various proportions without changing the basic structure. The Roman numerals indicate the valence state of atoms that often are found with more than one valence.

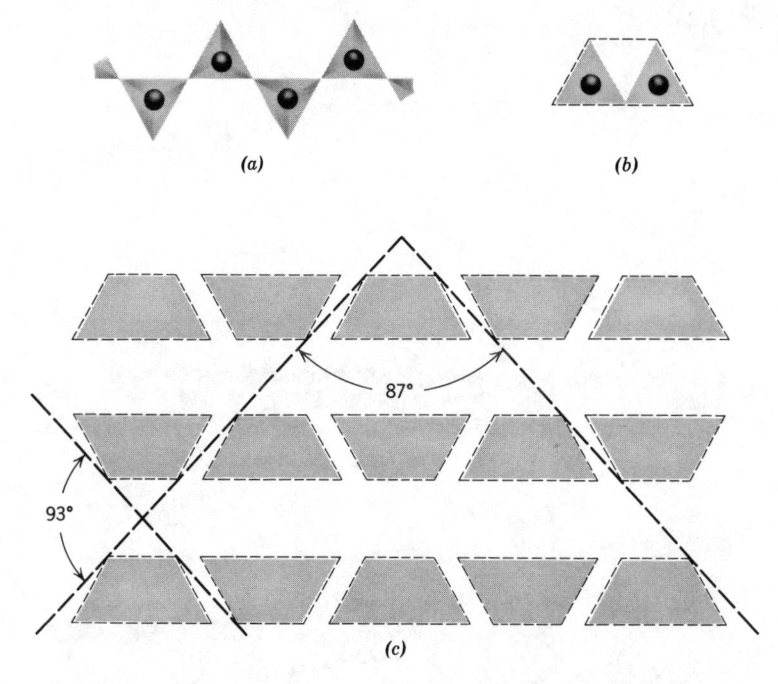

Figure 3.19 (*a*) The arrangement of silicate tetrahedra in one chain of a single-chain silicate. (*b*) A single chain viewed on end, showing the approximately trapezoidal cross-section. (*c*) The arrangement of single chains in a crystal, showing the angles of intersection of cleavage planes.

If, on the average, two and one-half corners of each tetrahedron are shared with each other, "double chains" of tetrahedra result, as shown in Figure 3.20. The ratio of oxygen to silicon is 11/4 to 1, making the formula of the silicate radical $(Si_4O_{11})^{-6}$. An example is the mineral, tremolite, $Ca_2Mg_5(OH)_2(Si_4O_{11})_2$. The arrangement of the double chains in a crystal accounts for the characteristic angles at which the cleavage planes intersect, 56° and 124°. In one form of tremolite, the dimensional mismatch between the $Ca_2Mg_5(OH)_2{}^{+12}$ sheets and the $(Si_4O_{11})_2{}^{-12}$ double chains causes the chains to roll into tight tubes, which are familiar as the fibers of one kind of asbestos.

When three out of four oxygens in every tetrahedron are shared, "sheet" structures with a silicate radical, $(Si_2O_5)^{-2}$, result (see

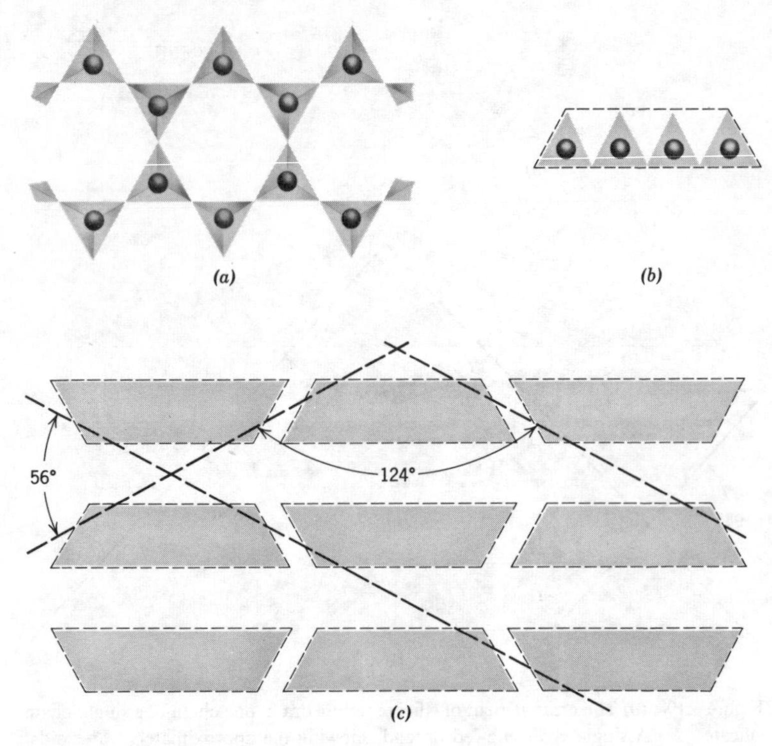

Figure 3.20 (*a*) The arrangement of silicate tetrahedra in one double chain of a double-chain silicate. (*b*) A double chain viewed on end, showing the approximately trapezoidal cross section. (*c*) The arrangement of double chains in a crystal, showing the angles of intersection of cleavage planes.

Figure 3.21). Crystals of kaolinite (a clay): $Al_2(OH)_4(Si_2O_5)$, talc: $Mg_3(OH)_2(Si_2O_5)_2$, and the micas: $KAl_2(OH)_2(AlSi_3O_{10})$ or $K(Mg, Fe^{II})_3(OH)_2(AlSi_3O_{10})$ are typical of the sheet silicate structure. The kaolinite crystal may be pictured as a $(Si_2O_5)^{-2}$ sheet bonded to an $Al_2(OH)_6$ sheet by replacing two-thirds of the $(OH)^-$ ions on *one* side of the $Al_2(OH)_6$ sheet by the unsatisfied oxygen atoms in the silicate sheet. The "one-sided" nature of the crystal helps to explain why clay absorbs water so readily; the polar kaolinite sheet attracts the polar water molecules by dipole forces. Talc does not absorb water so readily because the

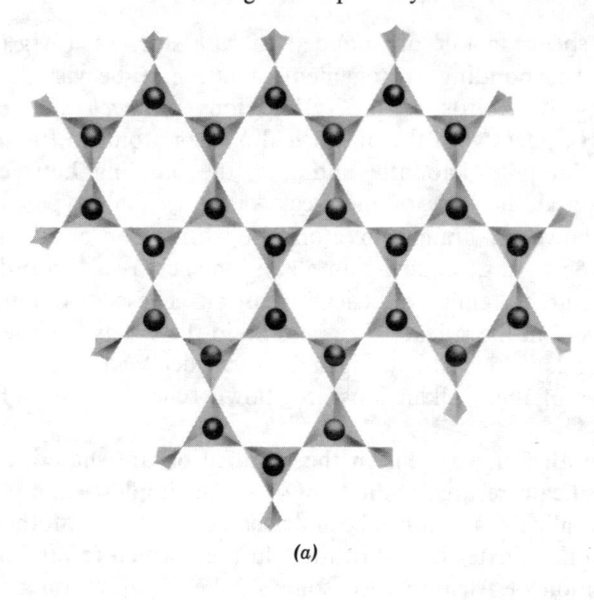

(a)

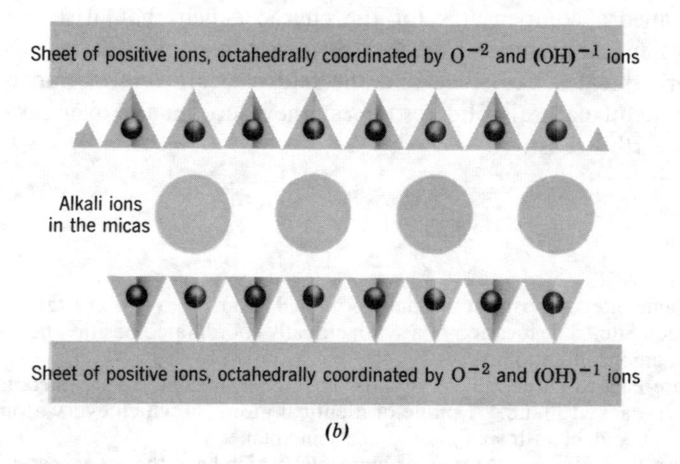

(b)

Figure 3.21 (*a*) The arrangement of silicate tetrahedra in a sheet silicate. (*b*) The arrangement of silicate sheets in a mica crystal, viewed on end. The alkali ions actually fit in the large holes shown in the view of (*a*). In talc and kaolinite the silicate tetrahedra are arranged similarly in a sheet, but there are no alkali ions holding the sheets together.

silicate sheets in talc are bonded to *both* sides of a $Mg_3(OH)_6$ sheet. The bonding arrangement in talc may be visualized by replacing two-thirds of the $(OH)^-$ ions on *each* side of the $Mg_3(OH)_6$ sheet with the unsatisfied oxygen atoms in the silicate sheets. In both kaolinite and talc, the bonding between the multilayered sheets is of the weak van der Waals type. In the micas, however, aluminum atoms substitute for one-fourth of the silicon atoms, requiring an alkali ion such as K^+ in order for the structure to remain electrically neutral. These alkali ions fit in the "holes" in the silicate sheet and bond the sheets together with ionic bonds, which are stronger than van der Waals bonds. The positions of these alkali ions are shown schematically in Figure 3.21.

When all four oxygens in the tetrahedron are shared, a "network" silicate results. Silica, SiO_2, is the simplest example of a network silicate, and it has been discussed already. Another class of network silicates is that of the feldspars, which result from the substitution of Al atoms for some of the Si atoms in the silica structure. The addition of alkali ions in the "holes" between the tetrahedra compensates for the charge deficit that arises when trivalent Al atoms are substituted for tetravalent Si atoms. Network silicates like silica and the feldspars are one of our most abundant natural solid resources; they account for over 50% of the earth's crust.

DEFINITIONS

Atomic Site: A position normally occupied by an atom in a crystal.

Cation Site: The location of a symmetrically coordinated position between anions in an ionic crystal.

Close-Packed Direction: A straight line along which atoms are in contact.

Close-Packed Plane: A plane of identical atoms in which every atom is in contact with six others in the same plane.

Close-Packed Sites: Atomic or ionic sites which have the same spatial arrangement as the atomic sites of one of the close-packed crystal structures, for example, face-centered cubic or hexagonal close-packed.

Crystal: A repetitive, three-dimensional arrangement of atoms or ions in a solid.

Crystal Structure: A mathematical representation of the relative positions of all atoms or ions in an ideal crystal.

Ionic Crystal: A crystal in which the predominant form of bonding is ionic.

Lattice Point: One point in an array, all the points of which have identical surroundings.

Lattice Translation: A vector connecting any two lattice points in the same lattice.

Molecular Crystal: A crystal in which the subunits associated with each lattice point are molecules.

Primitive Cell: A unit cell which has lattice points only at its corners; equivalently, a unit cell which contains only one lattice point.

Silicate: A mineral in which the subunits are $(SiO_4)^{-4}$ coordination tetrahedra; these tetrahedra may be bonded to one another by sharing oxygen atoms, or they may be bonded to positive ions in the structure by the remaining electron on each unshared oxygen atom.

Space Lattice: A three-dimensional array of points (lattice points) each of which has identical surroundings; also called a Bravais lattice.

Stacking Sequence: The manner in which close-packed planes are stacked on top of one another in a particular structure; planes stacked so their atoms are directly above one another are represented by the same letter, for example, ABABA . . . represents HCP stacking and ABCABCAB . . . represents FCC stacking.

Unit Cell: A convenient repeating unit of a lattice with lattice translations as its edges; a crystal structure unit cell is a unit cell which indicates atom positions as well as lattice points.

BIBLIOGRAPHY

INTRODUCTORY REFERENCES:

A. Holden and P. Singer, *Crystals and Crystal Growing,* Doubleday and Co., Garden City (1960).

A. E. H. Tutton, *The Natural History of Crystals,* E. P. Dutton and Co., New York (1924).

SUPPLEMENTARY REFERENCES:

D. H. Andrews and R. J. Kokes, *Fundamental Chemistry,* John Wiley and Sons, N. Y. (1962), Chapter 9.

L. V. Azároff, *Introduction to Solids,* McGraw-Hill Book Co., N. Y. (1960), Chapter 3.

C. Kittel, *Introduction to Solid State Physics,* John Wiley and Sons, N. Y. (1956), Chapter 1.

A. L. Loeb and G. W. Pearsall, "Moduledra Crystal Modules. A Teaching and Research Aid in Solid-State Physics," *American Journal of Physics,* Vol. 31 (1963) p. 190.

J. M. Robertson, *Organic Crystals and Molecules,* Cornell University Press, Ithaca (1953), Chapters II–III.

A. B. Searle and R. W. Grimshaw, *The Chemistry and Physics of Clays and Other Ceramic Materials,* Interscience Publishers, New York (1959), Chapters II–IV.

R. W. G. Wyckoff, *Crystal Strucures,* Sections I–IV, Interscience Publishers (1953); a reference work describing and classifying all known crystal structures of elements and compounds.

MORE ADVANCED TEXTS:

M. J. Buerger, *Elementary Crystallography,* John Wiley and Sons, New York (1956).
A. F. Wells, *Structural Inorganic Chemistry,* Clarendon Press, Oxford (1950).

PROBLEMS

3.1 Draw a hexagonal close-packed unit cell. Calculate the number of atoms per unit cell. Calculate the fraction of the total space occupied by spherical atoms in an HCP crystal.

3.2 Draw a hexagonal close-packed unit cell and a face-centered cubic unit cell. Identify both atom sites and lattice points in each.

3.3 Draw a cubic unit cell of NaCl. Identify the ion sites and the lattice points.

3.4 Give the coordinates of atom positions in cubic unit cells of MgO, K_2O and ZnS.

3.5 Determine the expected crystal structure, including ion positions, of the hypothetical salt AB_2 where the radius of A is 1.54 Å and the radius of B is 0.49 Å. Assume that A has a charge of +2.

3.6 Calculate the size of the largest sphere that could fit interstitially in a BCC crystal structure, as a fraction of the radius of the BCC atoms. The positions of the largest interstitial holes have coordinates of the type $0, \frac{1}{2}, \frac{1}{4}$ in a BCC unit cell.

3.7 Describe the crystal structure expected for CsCl. The coordination of Cl around Cs is eight-fold and both have a valence of 1. Compare this structure to that of CaF_2.

3.8 Compare the crystal structure of cubic ZnS to that of diamond.

3.9 Show that the structure of a brucite [$Mg(OH)_2$] sheet satisfies the valences of Mg, O, and H. This can be accomplished most easily by picking a repeating unit of the sheet and showing that it is neutral.

3.10 Calculate the values of x, y and m in $(Si_xO_y)^{-m}$ for

(a) single chain silicate
(b) double chain silicate
(c) sheet silicate
(d) network silicate

As in Problem 3.9, pick a repeating unit of the structure; show all calculations.

3.11 The hexagonal ring in the structure of cellulose contains a carbon atom at each of five corners and an oxygen atom in the sixth. Sketch the arrangement of carbon tetrahedra that would satisfy the 120° angles of the hexagon.

3.12 Explain the fact that talc is much softer than mica, although both show pronounced basal cleavage.

The following problems are based on the material presented in Appendix III as well as that in Chapter 3.

3.13 Calculate the theoretical c/a ratio for hexagonal close-packing. The following values have been measured for some common hexagonal metals and are here listed for comparison: Be: 1.58, Cd: 1.89, Mg: 1.62, Ti: 1.60, Zn: 1.85, Zr: 1.60. See Figure III.3 for c and a dimensions of the hexagonal unit cell.

3.14 Determine the Miller indices of the family of close-packed planes in a face-centered cubic crystal. Determine the Miller indices of the family of close-packed directions in a face-centered cubic crystal. Determine the Miller indices of the family of close-packed directions in a body-centered cubic crystal.

3.15 Sketch the following planes and directions in cubic unit cells: (123), [123], (112), [112], $(\bar{1}10)$, $[\bar{1}10]$. Show that each of the planes above contains the $[11\bar{1}]$ direction.

3.16 (b) Draw a (111) plane in a tetragonal unit cell.
(b) Draw a (110) plane in an orthorhombic unit cell.
(c) Draw a (100) plane in a monoclinic unit cell.

3.17 What are the Miller indices of the line of intersection of a (111) plane and a $(11\bar{1})$ plane in a cubic crystal? Determine the answer both by sketching the planes and their intersection and by using the cross product of the two normals.

3.18 Consider atoms to be hard spheres in contact and calculate, for face-centered cubic, body-centered cubic and simple cubic packing, the following:
(a) the volume of the cubic unit cell in terms of an atom diameter D
(b) the number of atoms per unit cell
(c) the density in terms of number of atoms/D^3

The following problem is based on the material presented in Appendix IV as well as in Appendix III and Chapter 3.

3.19 Intensity peaks were measured at the following values of θ for a powdered sample of cubic metal bombarded with Cu radiation having a wavelength $\lambda = 1.54$ Å: 20°, 29°, 36.5°, 43.4°, 50.2°, 57.35°, 65.55°. Calculate the lattice parameter and determine the crystal structure of the metal. Identify the metal by referring to tabulated data for the elements, as may be found for example in the *Handbook of Chemistry and Physics.*

CHAPTER FOUR

Imperfections in Crystals

Crystals are rarely perfect. Many of the important properties of crystalline materials are determined by various imperfections in them. The imperfections that can be described as disruptions in the space lattice are called lattice imperfections and can be characterized geometrically, according to whether the center of the disruption is at a point, along a line, or over a surface.

4.1 INTRODUCTION

Real crystals usually are not composed simply of identical atoms on identical sites throughout a regularly repeating three-dimensional lattice. They contain imperfections or defects. The imperfections which disrupt the structure most are the imperfections in the space lattice. Since the space lattice is a geometric concept, it is natural to classify lattice imperfections geometrically. Thus, zero-dimensional (point), one-dimensional (line), and two-dimensional (surface) imperfections are treated separately. Volume imperfections, such as the thermal excitation of all the atoms off their lattice sites, exist but will not be covered here (see Volume IV, *Electronic Properties*). Also, imperfections in electronic structure of atoms will not be treated, for the emphasis of the chapter is on defects in atomic arrangements.

4.2 POINT IMPERFECTIONS

A point imperfection is a very localized interruption in the regularity of a lattice. A point imperfection comes about, as a rule, because of the absence of a matrix atom (an atom that would be present in a perfect crystal), the presence of an impurity atom, or a matrix atom in the "wrong" place (a site not occupied in the perfect crystal). The most common point imperfections in a crystal of a pure element are illustrated in Figure 4.1; some of those that may occur in an ionic crystal are shown in Figure 4.2. The absence of an atom from a normally occupied site is called a *vacancy.* A foreign atom which is in the site of a matrix atom is called a *substitutional impurity atom,* and a foreign atom in an interstice between matrix atoms is called an *interstitial impurity atom.* There are, in addition, point imperfections which are more complex, for example, condensed groups of vacancies such as *di-vacancies* and *tri-vacancies.* In a crystal, an atom may leave its site, creating a vacancy, and dissolve interstitially in the structure. The associated vacancy and interstitial atom is called a *Frenkel imperfection.* When a cation vacancy is associated with an anion vacancy rather than an interstitial cation, the pair (the anion vacancy and the associated cation vacancy) is called a *Schottky imperfection.* In general, Schottky imperfections are more likely to form than Frenkel imperfections, for few structures contain large enough interstices to

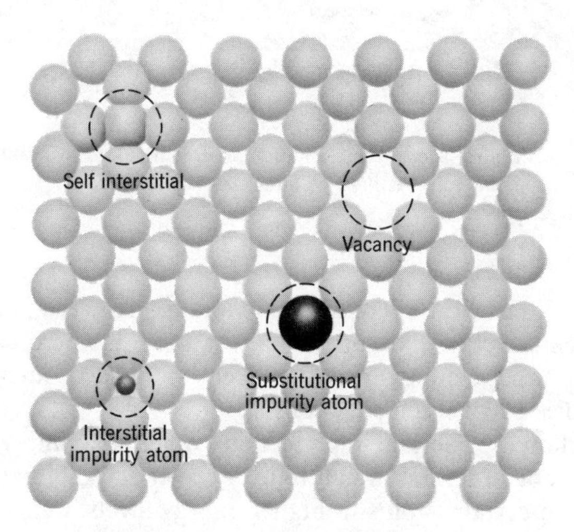

Figure 4.1 A two-dimensional representation of a simple crystalline solid, illustrating some of the point imperfections that are possible.

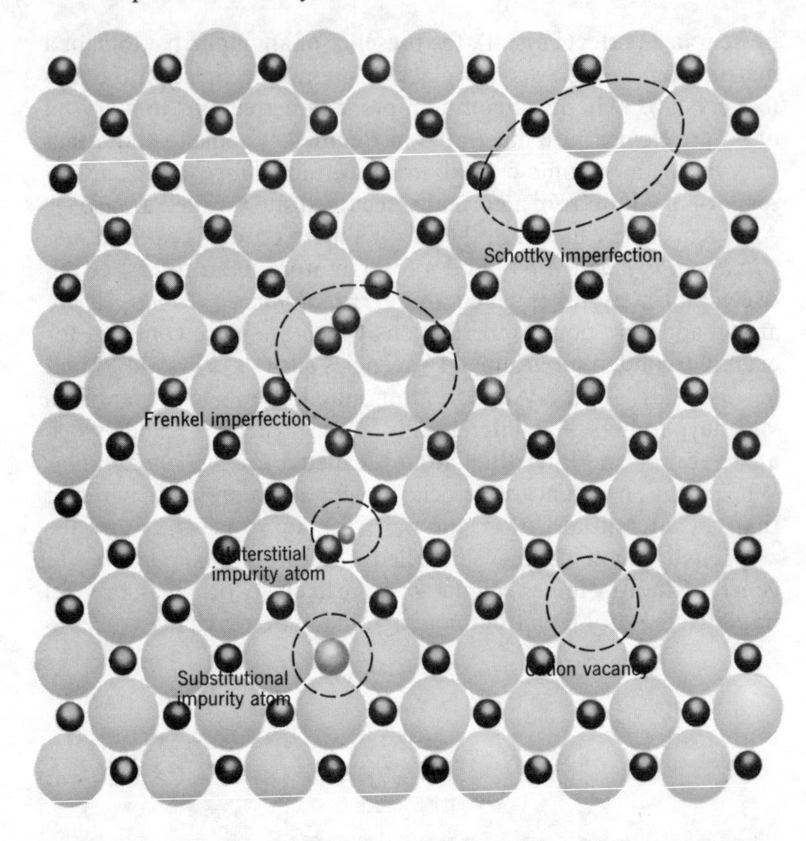

Figure 4.2 A two-dimensional representation of an ionic solid, illustrating a number of possible point imperfections. Cations are represented by small spheres, anions by large spheres.

dissolve cations without considerable strain. However, in a relatively open structure like that of fluorite (CaF_2), where the interstices are large, cations are quite likely to dissolve interstitially and produce Frenkel imperfections.

In addition to structural imperfections, other point imperfections which are not structural in nature may be produced by raising the electrons in certain atoms to higher energy levels. This may be accomplished by bombarding the crystal with electrons (resulting in X-rays), microwave radiation (the principle of the maser), light (lasers and photoelectric materials), and other forms of radiation of suitable frequency. These will be discussed in Volume IV, *Electronic Properties.*

4.3 LINE IMPERFECTIONS

Line imperfections, like point imperfections, are defined by the way in which their presence causes disruptions in what otherwise would be a perfect space lattice. As the name implies, the distortion around a line imperfection is centered along a line; thus the imperfection can be considered as the boundary between two regions of a surface which are perfect themselves but are out of register with each other. For example, in Figure 4.3*a* a block of an ideal simple cubic lattice is pictured; for clarity the lattice points (intersections of the grid lines) are shown only for the two adjacent planes on either side of the midplane in the block. Now, if the lattice points in the central portion of the upper plane are displaced one-half lattice translation to the left and the lattice points in the central portion of the lower plane are displaced one-half lattice translation to the right, the lattice is distorted as shown in Figure 4.3*b*. In this figure we have cut the block in four and separated the pieces in order to emphasize the details of the lattice distortion. The same sequence is repeated in Figure 4.4 but here the block is viewed from above, looking down on the midplane. The view in Figure 4.4*b* shows that the lattice is in perfect register both in the central portion where the lattice points have been displaced from their initial positions and in the outer portion where there has been no relative displacement; only at the boundary between the two regions is the lattice distorted. We say that the central portion has *slipped* and the outer portion has not, and we call the midplane, across which slip has taken place, the *slip plane.* The line imperfection which is the boundary between the "slipped" region and the "unslipped" region lies in the slip plane and is called a *dislocation.* The structure and behavior of dislocations are fundamental to understanding many of the properties of engineering materials. Their structure will be emphasized in the following paragraphs, and their behavior will be treated in further detail in Volume III.

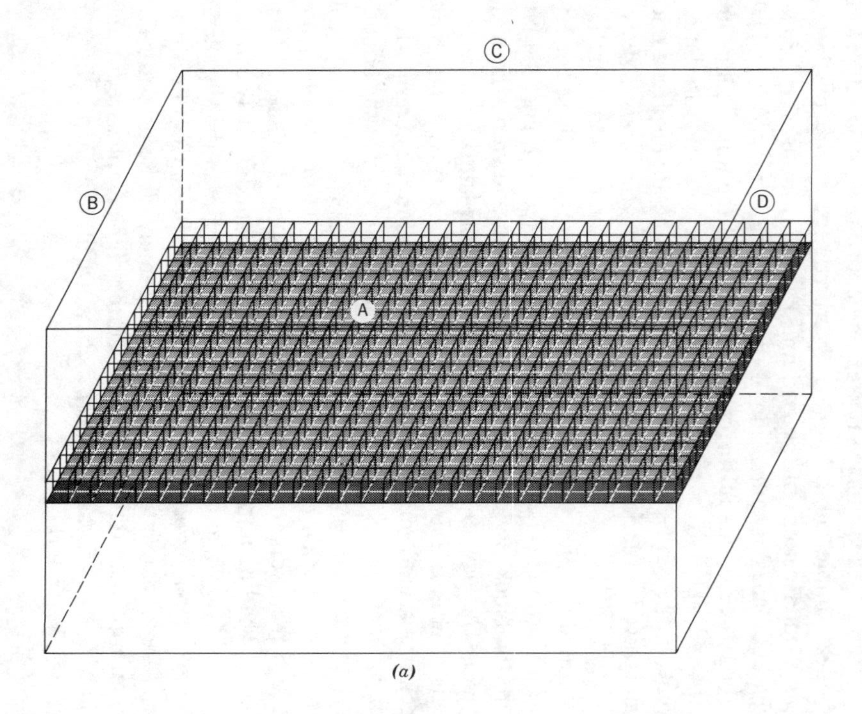

Figure 4.3 (*a*) A block of an ideal simple cubic lattice. Lattice points are shown (as line intersections) only for two planes in the middle of the block. (*b*) The lattice distortion that results when the lattice points in the central part of one plane are displaced one-half lattice translation to the left while those in the plane below are displaced one-half lattice translation to the right is shown below. The block of Figure 4.3*a* has been cut in four after the deformation in order to show the local distortion more clearly.

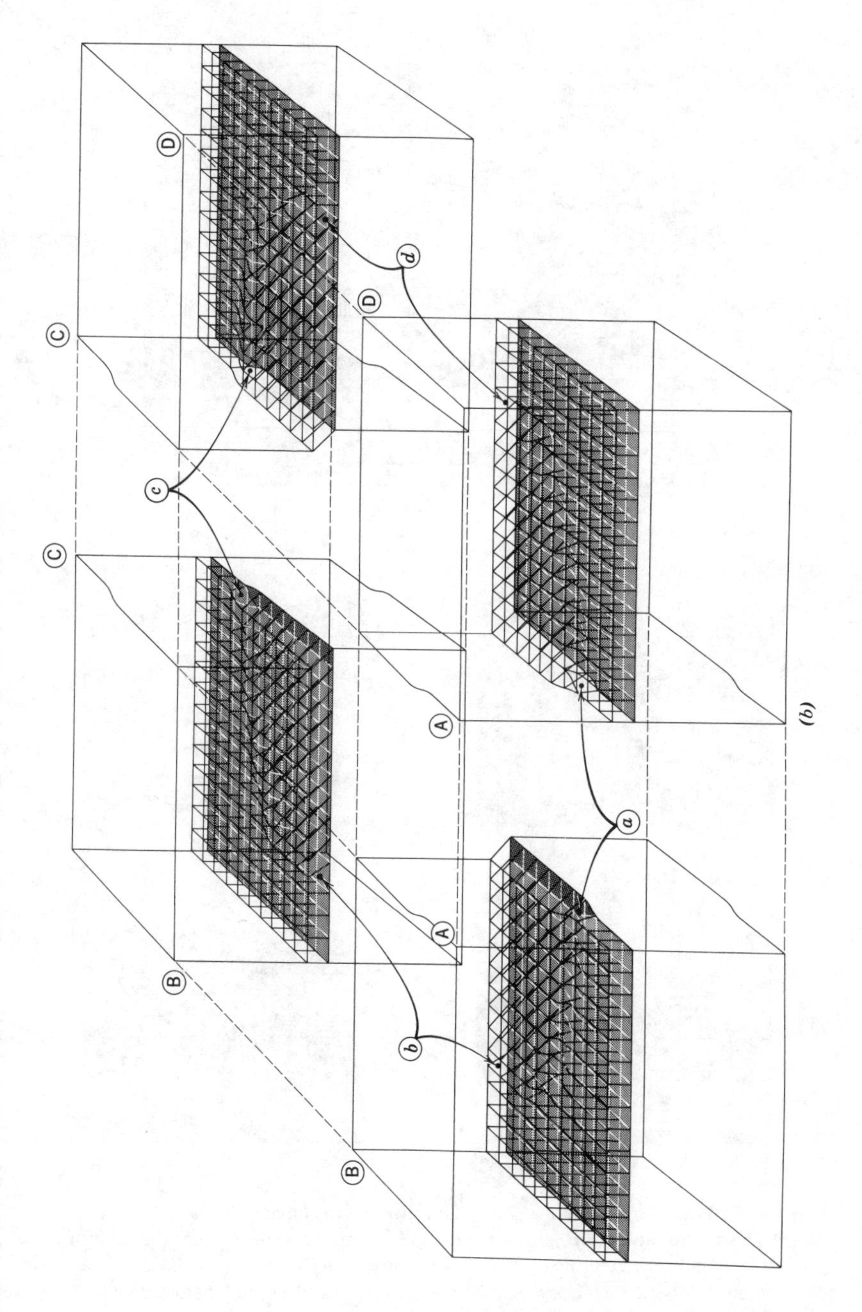

(b)

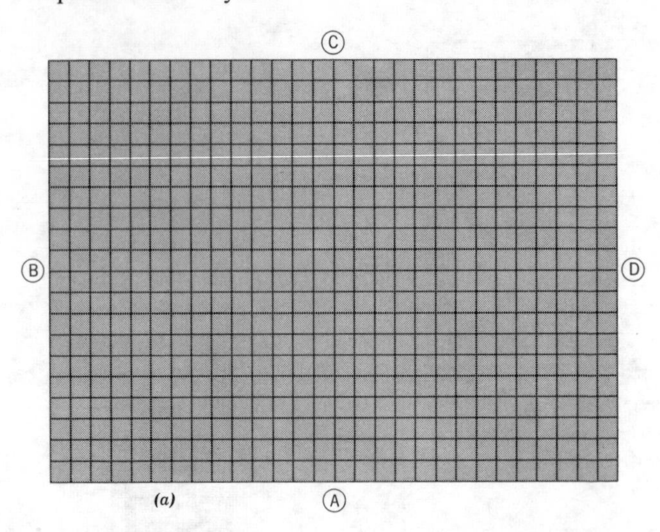

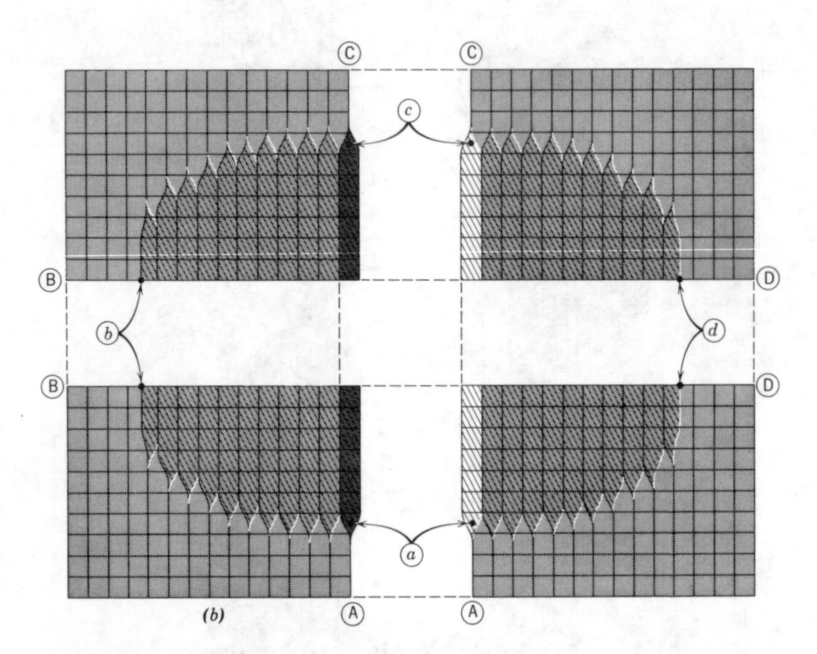

Figure 4.4 (*a*) The block of Figure 4.3*a* viewed from above. The lattice is perfect. (*b*) The cut-up block of Figure 4.3*b* viewed from above, showing that the lattice is perfect both in the central region (cross-hatched) and in the outer region. Only at the boundary between the two regions, the *dislocation,* is the lattice distorted severely.

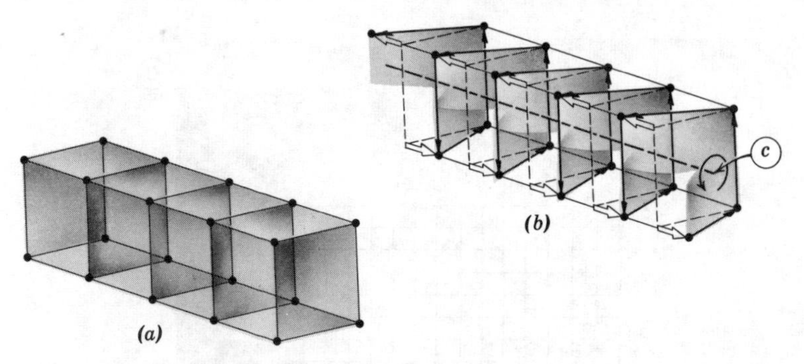

Figure 4.5 (*a*) An array of four-unit cells in an ideal simple cubic lattice. (*b*) The lattice points of Figure 4.5*a* after having been displaced by the broad arrows, resulting in a screw dislocation along the heavy dashed line. The lighter arrows trace a helical path through the distorted lattice around the dislocation. Point c refers to Figure 4.3*b*.

In Figures 4.3*b* and 4.4*b* it can be seen that the nature of the lattice distortion produced by a dislocation is a function of position around the loop. For example, looking left down the dislocation line at point *c* in the block, would show that the lattice points next to the dislocation could be pictured as lying on a helix (see Figure 4.5). The same is true at point *a* but the helix is of opposite sense. For this reason the dislocation segments at points *a* and *c* are described as *screw* dislocations (symbol: ↻ or ↺). On the other hand, looking back along the dislocation line at point *b* in the block, the lattice appears to be distorted by squeezing in an extra half-plane of lattice points above the slip plane (see Figure 4.6); similarly, at point *d* the distortion can be described in terms of an extra half-plane squeezed in below the slip plane. In such a case, the dislocation segment is called an *edge* dislocation (symbol: ⊥ or ⊤) because it appears to lie at the edge of the extra half-plane. An edge dislocation also differs from a screw dislocation in the type of strain the lattice experiences around it. Figure 4.7 illustrates the fact that the lattice is in compression on the side of the slip plane where the extra half-plane is and in tension on the other side. In contrast, a screw dislocation has neither tensile nor compressive strains, but rather shear strains,[1] associated with it.

[1] The distinctions between tensile, compressive, and shear strains are discussed in Volume III of this series.

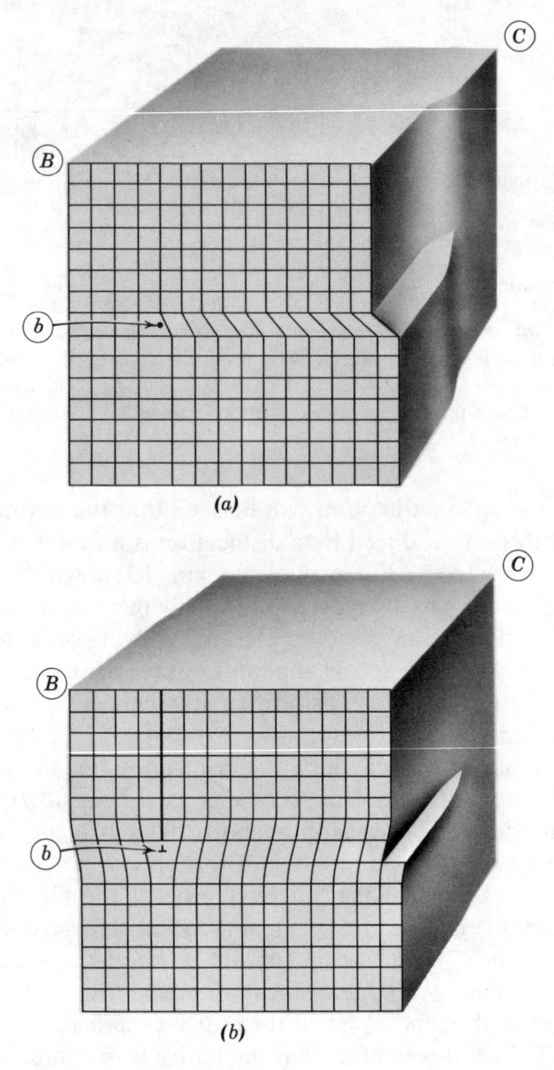

Figure 4.6 (*a*) The lattice distortion on one cut face of the block in Figure 4.3*b*. (*b*) The block of Figure 4.6*a* after allowing the lattice points to relax to more-or-less equilibrium positions. The edge of the extra half plane is drawn as a heavy line and the edge dislocation intersects the front surface at point *b*.

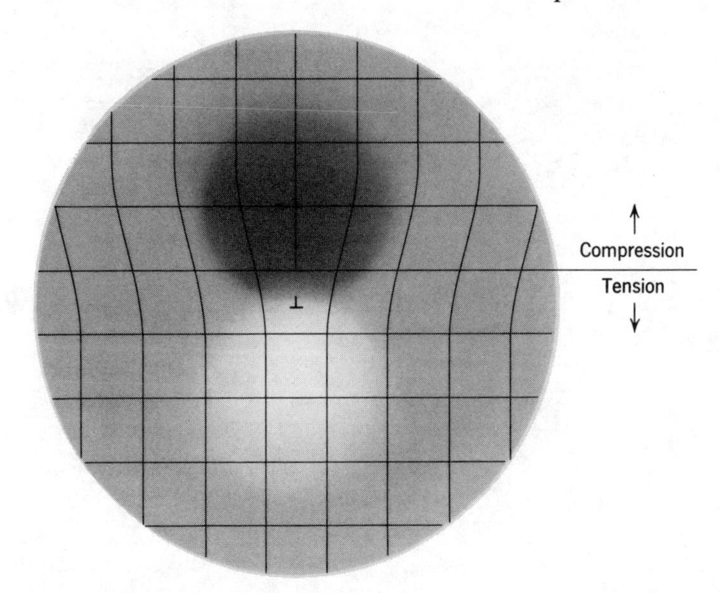

Figure 4.7 The regions of tension (light) and compression (dark) around an edge dislocation in a simple cubic lattice.

Dislocations may also be described in terms of two vectors. The direction of the dislocation *line* at any given point is described by the unit vector, **t** (Figure 4.8), which is parallel to the line and must be continuous around a dislocation loop, that is, it will have opposite directions on opposite sides of a loop. Whether it is assigned in a clockwise or a counterclockwise direction is arbitrary, resulting in a plus-or-minus ambiguity. The second vector is the translation vector (more commonly called the *Burgers vector*) **b,** which indicates how much and in what direction the lattice above the slip plane appears to have been shifted with respect to the lattice below the slip plane (Figure 4.8). Note that the Burgers vector is exactly the same at any point on the loop, whereas the vector **t,** changes direction continuously. The Burgers vector of any dislocation can be determined by using the fact that a series of lattice vectors that form a closed circuit in a perfect lattice will not form a closed circuit if they surround a dislocation; the vector necessary to close the circuit around the dislocation is the Burgers

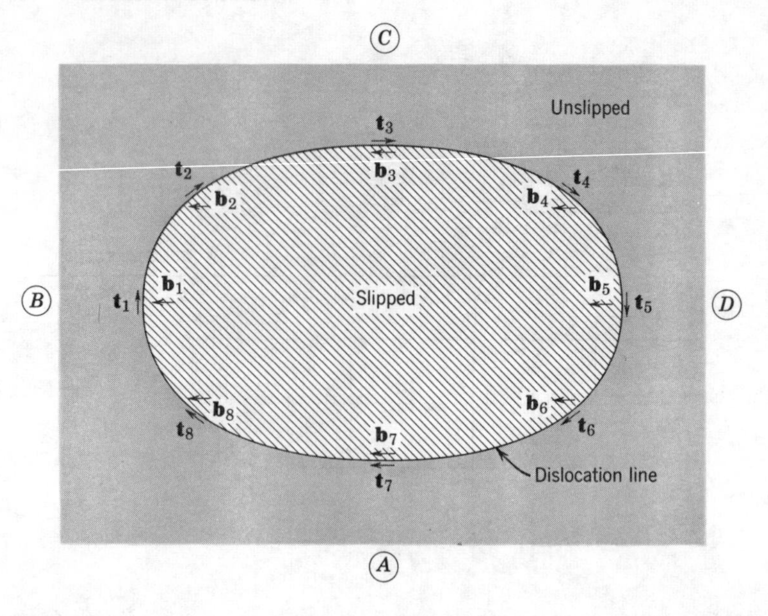

Figure 4.8 A schematic presentation of the dislocation loop shown in Figure 4.4*b*. The unit vector of the dislocation *line* is designated by the vector **t** at various points around the loop. The Burgers vector is constant around the loop and is designated by the vector **b**.

vector of the dislocation. Two such *Burgers circuits* around the same dislocation are shown in Figure 4.9.

We are now in a position to redefine *edge* and *screw* dislocation segments. A dislocation segment is edge when **b** is perpendicular to **t**, and a dislocation segment is screw when **b** is parallel to **t**. When the angle between **b** and **t** is between 0 and 90°, the dislocation is called *mixed*.

The Burgers vector of a dislocation in a particular structure is described by the Miller indices (Appendix III) of the direction and a fraction indicating the amount by which each direction component is to be multiplied. For example, in FCC crystals the Burgers vector is one lattice translation in a $\langle 110 \rangle$ type direction, as indicated by the Burgers circuit in the close-packed plane in Figure 4.10. Since a $\langle 110 \rangle$ unit vector represents two lattice translations, $\mathbf{b} = \frac{1}{2}\langle 110 \rangle$ for FCC crystals.

The Burgers vectors considered so far have all been lattice translations. They connected one lattice point to a neighboring one. However, when we start to deal with the fact that, in crystals, atoms are associated with the lattice points, it becomes evident that positions of relative equilibrium may occur between lattice points. Such is the case in FCC crystals. Figure 4.11 shows two adjacent $(1\bar{1}1)$ atomic planes in a FCC crystal. The Burgers vector, $\overline{ab}$, is $\frac{1}{2}[110]$. But an atom moving directly from a to b would have to ride up over an atom in the plane below; instead we might expect it to go "through the valley" to c and then from c to b. The vectors $\overline{ac}$ and $\overline{cb}$ are the Burgers vectors of *partial dislocations.* Their representation in Miller indices is $\overline{ac} = \frac{1}{6}[21\bar{1}]$ and $cb = \frac{1}{6}[121]$; they add vectorially to $\frac{1}{2}[110]$.

The passage of a dislocation through a crystal results in the relative motion of one part of the crystal past the other part. When a dislocation moves in a plane containing both the line vector **t** and the Burgers vector **b**, the process is called *slip.* The slip process is illustrated in Figure 4.12 for a pure edge dislocation and a pure screw dislocation, although, in general, slip is more likely to occur by the expansion of a dislocation loop on a slip plane. If

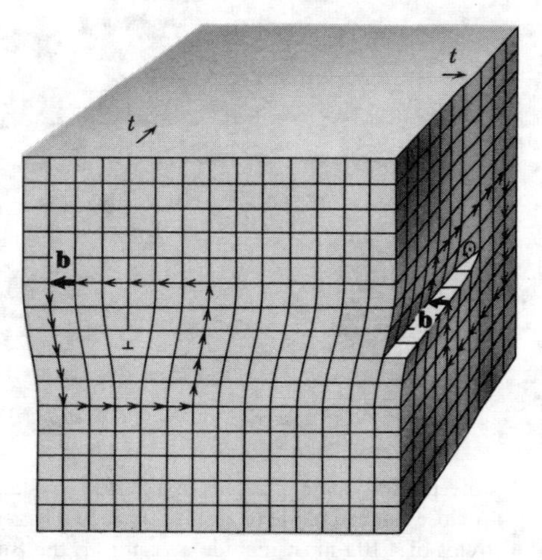

Figure 4.9 Two Burgers circuits, drawn around different segments of the same dislocation. Each circuit is 5 x 5 and would close in a perfect lattice but here each circuit must be completed by a vector, **b**, the Burgers vector of the dislocation.

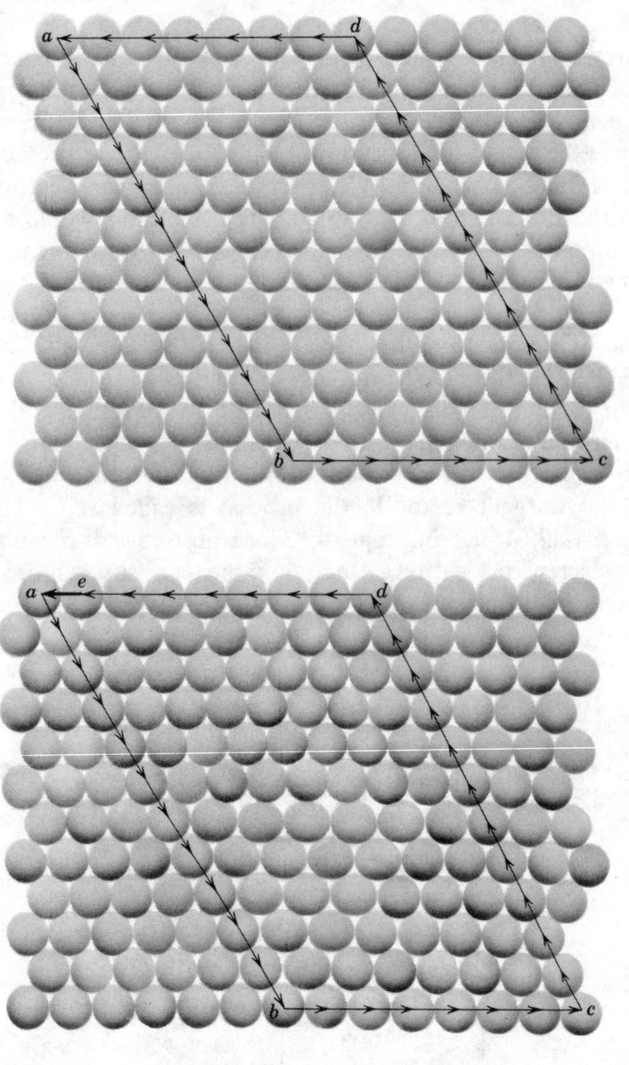

Figure 4.10 (*a*) A close-packed plane of atoms in a perfect crystal, showing a circuit that closes. (*b*) A close-packed plane of atoms intersected by a dislocation. In order to close, the circuit of 4.10*a* must include a vector $\overline{ae}$, the Burgers vector of the dislocation. In this case, the Burgers vector is in a direction of the type $\langle 110 \rangle$.

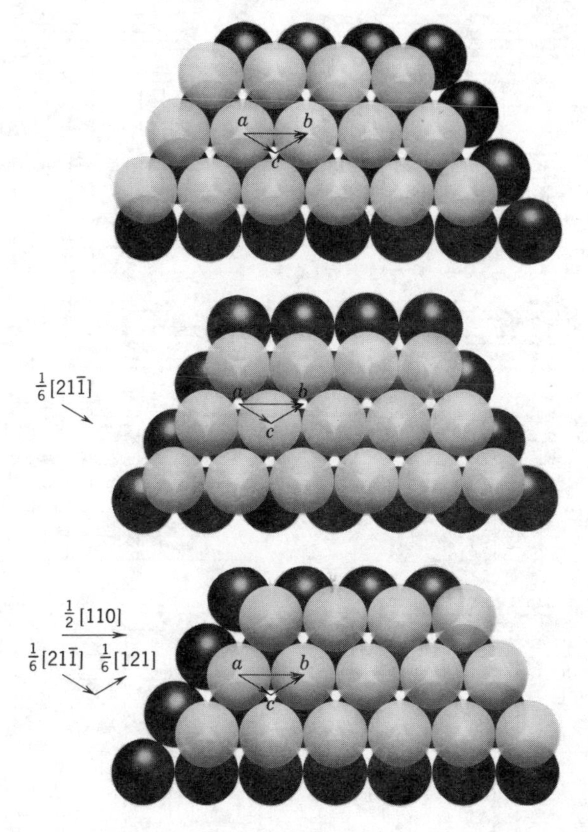

Figure 4.11 The zig-zag motion of one (111) plane over another in a FCC crystal, illustrating the meaning of partial dislocations. The sum of the Burgers vectors of the two partials, $\frac{1}{6}[21\bar{1}]$ and $\frac{1}{6}[121]$, is a lattice translation vector, that is, the Burgers vector of a whole dislocation.

the dislocation moves in any other plane, the process is called *climb.* The details of this dislocation movement will be discussed in Volume III, *Mechanical Behavior.*

4.4 SURFACE IMPERFECTIONS

Surface imperfections of a structural nature arise from a change in the stacking of atomic planes across a boundary. The change

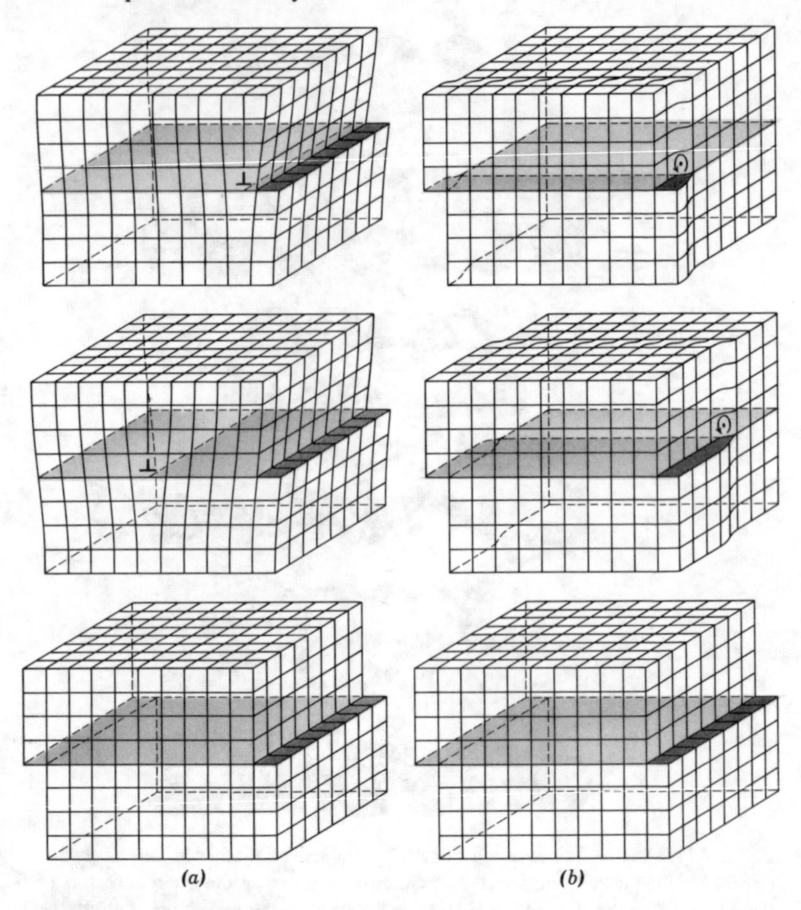

Figure 4.12 (*a*) Slip resulting from the movement (right to left) of a pure edge dislocation through a simple cubic lattice. (*b*) Slip resulting from the movement (front to back) of a pure screw dislocation through a simple cubic lattice. The dislocations in (*a*) and (*b*) have the same Burgers vector so they result in the same amount and direction of slip.

may be one of the orientation or of the stacking sequence of the planes.

Grain boundaries are those surface imperfections which separate crystals of different orientations in a polycrystalline aggregate. As illustrated for a two-dimensional analogy in Figure 4.13, the boundary atoms in two randomly orientated grains cannot have a

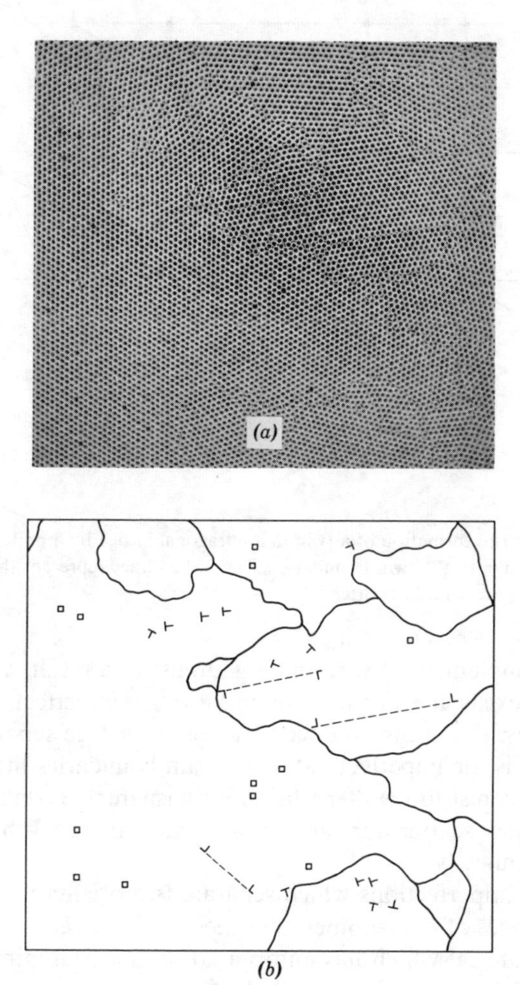

Figure 4.13 (*a*) A photograph showing two-dimensional crystalline regions in an array of soda straws. The grain boundaries and other defects appear more clearly if the photograph is viewed from a small angle to the page. (Photograph courtesy of G. Falla.) (*b*) A schematic representation of Figure 4.13*a*. Grain boundaries are indicated by lines, dislocations by $\perp$, extended dislocations by ⅃– – – – – – –L and vacancies by □.

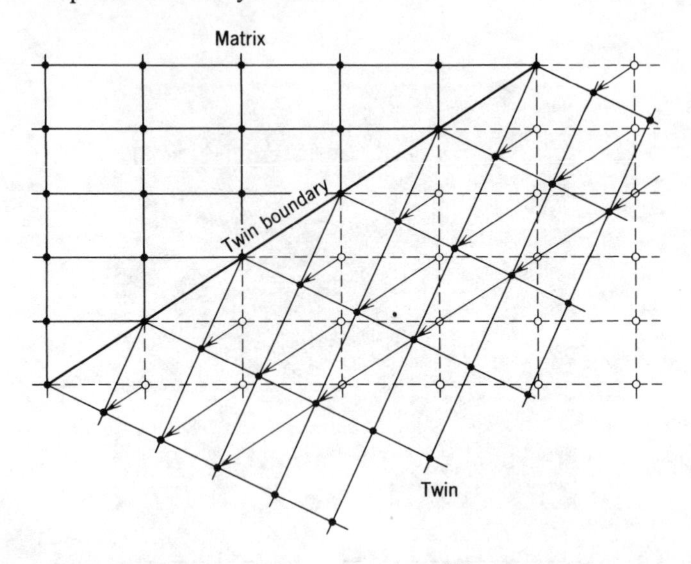

Figure 4.14 The formation of a twin in a tetragonal lattice by a uniform shearing of atoms parallel to the twin boundary. The dashed lines represent the lattice before twinning, the solid lines after.

perfect complement of surrounding atoms; as a result, a region of transition exists in which the atomic packing is imperfect. In three-dimensions, this transition occurs across a surface separating the grains. It is the imperfect nature of grain boundaries that enables the microscopist to see them, for in a transparent crystalline material they may scatter light and in an opaque material they can be etched chemically.

Surface imperfections which separate two orientations that are mirror images of one another are called *twin boundaries.* The volume of material which has an orientation that is a mirror image of the matrix orientation is called a *twin.* Twins may come into existence during the growth of a crystal or they may arise during deformation. Shear parallel to the twin boundary can produce a twin, as shown in Figure 4.14, particularly if slip is difficult to initiate or propagate in that direction. The lattice shear is uniform, that is, the displacement of a lattice point in the twinned region is directly proportional to the distance from the boundary and parallel to the twin boundary. If more than one atom is associated with each lattice point, as in HCP metals, minor atomic readjustments may have to occur after the twinning shear if the crystal structure as well as the lattice is to be twinned.

A *stacking fault* is a surface imperfection that results from the stacking of one atomic plane out of sequence on another while the lattice on either side of the fault is perfect. For example, the stacking sequence in an ideal FCC crystal may be described as ABCABCABC.... A stacking fault might change the sequence to ABCABABCA .... The stacking fault in this case is due to the "A" plane of atoms after the second "B" and may be described as a very thin region of HCP stacking in a FCC crystal. Such stacking faults may occur during crystal growth or may result from the separation of two partial dislocations. In either case, the crystalline material on one side of the imperfection has the same orientation as that on the other side but is translated with respect to it by a fraction of a lattice vector (in Figure 4.15 by the Burgers vector of a partial dislocation).

Another surface imperfection is a *low-angle boundary,* which is really a limiting case of a grain boundary where the angular misorientation is of the order of a few degrees. In general, low-angle

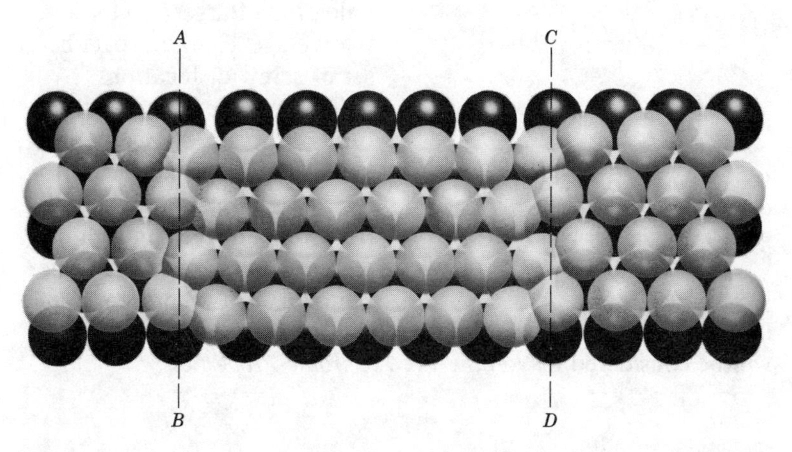

Figure 4.15 A view normal to a stacking fault between the lines *AB* and *CD* on a close-packed plane, in either a FCC or a HCP metal. If the stacking fault is viewed as resulting from the dissociation of a dislocation into two partials, *AB* and *CD* are the lines of the partial dislocations.

boundaries can be described by suitable arrays of dislocations. A low angle *tilt* boundary is composed of edge dislocations lying one above the other in the boundary (see Figure 4.16). The angle of tilt will be

$$\theta = \frac{b}{h} \tag{4.1}$$

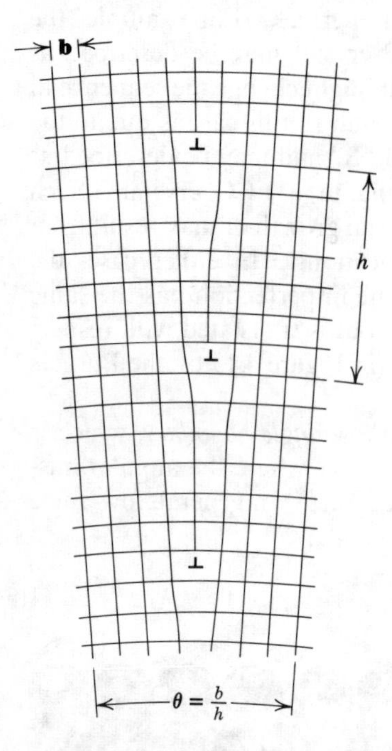

Figure 4.16 A simple tilt boundary composed of edge dislocations a distance h apart in a simple cubic lattice.

where b is the magnitude of the Burgers vector and h is the average vertical distance between dislocations. A low-angle *twist* boundary (Figure 4.17) can be described in terms of at least two sets of parallel screw dislocations lying in the boundary. In this case, the angle of twist is

$$\alpha = \frac{b}{h} \tag{4.2}$$

where b and h are the magnitude of the Burgers vector and the average separation of either set of screw dislocations.

Of course, many nonstructural surface imperfections also exist. For instance, when two ferromagnetic regions differ from one another only in the direction of magnetization, the boundary between them is an imperfection and is called a ferromagnetic domain wall. These will be considered in Volume IV, *Electronic Properties.*

DEFINITIONS

Burgers Circuit: A sequence of connected unit lattice translation vectors forming a circuit which would close in a perfect lattice but which fails to close when taken around a dislocation. The vector necessary to

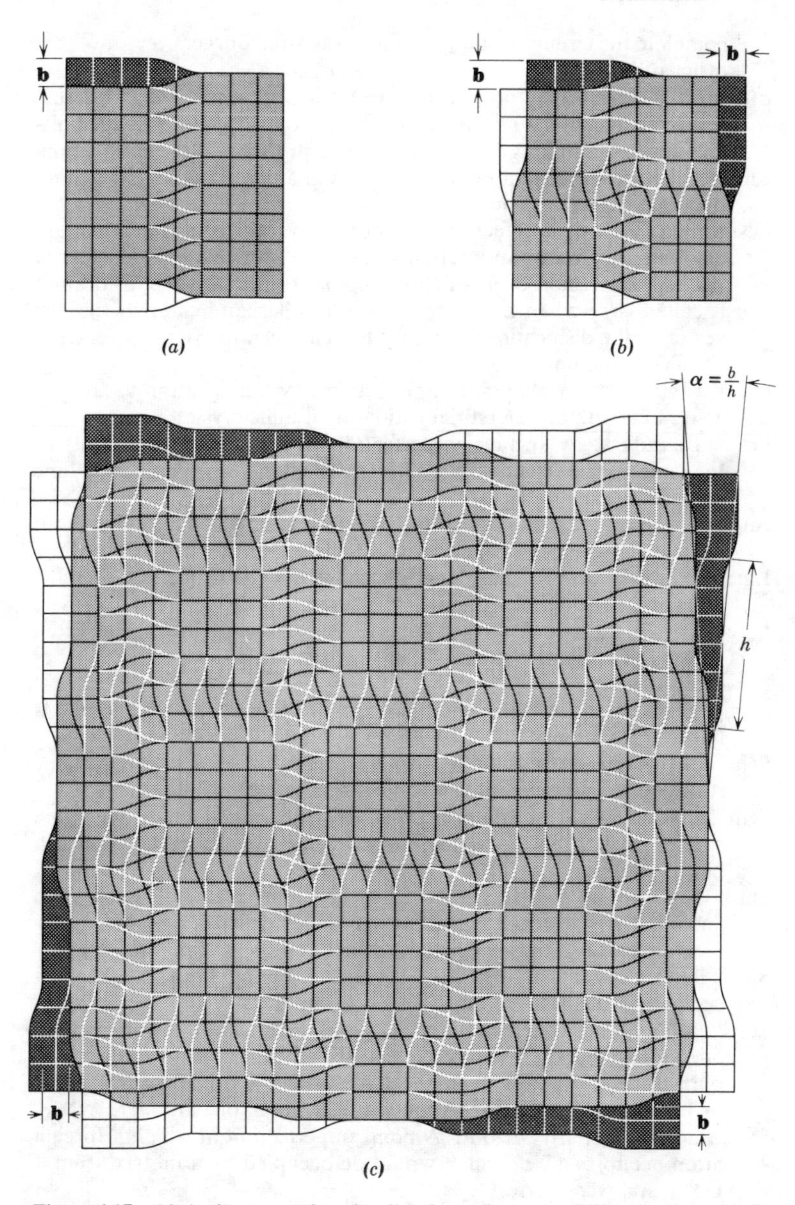

Figure 4.17 (*a*) A view normal to the slip plane of a screw dislocation in a simple cubic lattice. (*b*) Two intersecting screw dislocations in a simple cubic lattice. (*c*) A simple twist boundary composed of two sets of intersecting screw dislocations in a simple cubic lattice.

complete the circuit around a dislocation is the Burgers vector of that dislocation.

Burgers Vector: The vector by which the lattice on one side of an internal surface is displaced relative to the lattice on the other side as a dislocation moves along the surface; it is a property of the dislocation.

Climb: The movement of a dislocation along any internal surface other than one of its slip planes.

Dislocation: A line imperfection which can be visualized as the boundary between a region of an internal surface over which slip has occurred and another region over which no slip has occurred. The dislocation is called *edge* when the Burgers vector is perpendicular to the line vector of the dislocation; it is called *screw* when the Burgers vector is parallel to the line vector.

Frenkel Imperfection: A point imperfection in which a cation vacancy is associated with an interstitial cation in an ionic crystal.

Grain Boundary: A surface imperfection which separates crystals of the same crystal structure but different orientations in a polycrystalline aggregate.

Interstitial Impurity Atom: A point imperfection in which a foreign atom fits in an interstice between matrix atoms.

Line Vector: A vector of unit length which is parallel to the line of a dislocation at any point along it.

Low-Angle Boundary: A surface imperfection separating two misoriented regions of a crystal; the angle of misorientation is small (a few degrees or less).

Partial Dislocation: A dislocation having a Burgers vector which is *not* a lattice translation.

Schottky Imperfection: A point imperfection in which a cation vacancy is associated with an anion vacancy in an ionic crystal.

Self-Interstitial: A point imperfection in which an atom of the same species as those in the matrix is squeezed into an interstice between matrix atoms.

Slip: The sliding displacement of one part of a crystal relative to another by the motion of a dislocation or dislocations; the movement of a dislocation in one of its slip planes.

Slip Plane: Any crystallographic plane containing both the Burgers vector and the line vector of a dislocation.

Stacking Fault: A surface imperfection which results from the stacking of one atomic plane on another out of sequence, so that the lattices on both sides of the fault have the same orientation but are translated by less than a lattice translation with respect to one another.

Substitutional Impurity Atom: A point imperfection in which a foreign atom occupies a site which would be occupied by a matrix atom if the crystal were perfect.

Tilt Boundary: A low-angle boundary in which the misorientation is a rotation about an axis lying in the boundary.

Twin Boundary: A surface imperfection separating two regions of a crystal which are mirror images of each other with respect to the plane of the boundary.

Twist Boundary: A low-angle boundary in which the misorientation is a rotation about an axis normal to the boundary.

BIBLIOGRAPHY

INTRODUCTORY REFERENCES:

W. C. Dash and A. G. Tweet, "Observing Dislocations in Crystals," *Scientific American,* Vol. 205 (October 1961), p. 107.

N. F. Mott, *Atomic Structure and the Strength of Metals,* Pergamon Press, New York (1956).

SUPPLEMENTARY REFERENCES:

Dislocations in Metals, ed. by M. Cohen, *AIME,* New York (1954).

C. Kittel, *Introduction to Solid State Physics,* John Wiley and Sons, New York (1956), Chapters 17 and 19.

MORE ADVANCED TEXTS:

A. H. Cottrell, *Dislocations and Plastic Flow in Crystals,* Clarendon Press, Oxford (1953).

W. T. Read Jr., *Dislocations in Crystals,* McGraw-Hill Book Co., New York (1953).

H. G. Van Bueren, *Imperfections in Crystals,* Interscience Publishers, New York (1960).

PROBLEMS

4.1 The larger the coordination around an interstice, the larger is its size relative to the size of the coordinating atoms. Using this fact, explain why Frenkel imperfections are more likely in fluorite (CaF_2) than in lithium fluoride (NaCl crystal structure).

4.2 When substitutional atoms do *not* occupy random sites in a crystal structure, an *ordered structure* results. For example, even though Cu and Au show complete solid solubility at elevated temperatures, at lower temperatures ordering takes place both at about 25 atom per cent Au and 50 atom per cent Au. Sketch a FCC unit cell, and designate the most likely sites for the Au atoms in Cu_3Au (such that as much symmetry as possible is preserved in the structure). Do the same for the 50 atom per cent Au ordered structure. Does the ordered or the disordered structure have the most point imperfections?

4.3 Calculate the largest diameter of an atom which could fit interstitially in a copper crystal without distorting it. The edge length of the FCC unit cell of Cu is 3.61 Å.

4.4 Describe how the movement of atoms in a crystal might be accomplished by the motion of vacancies. Could the same result be achieved by the motion of self-interstitials?

4.5 Show that vacancies must either be created or annihilated during the climb of a *pure edge* dislocation in a direction perpendicular to its slip plane.

4.6 Explain why an undissociated (not separated into partials) *pure screw* dislocation does not climb, but always moves by slip.

4.7 Would you expect a dislocation to dissociate on the close-packed plane of a HCP crystal? Why or why not?

4.8 Give three differences between dislocations in a simple cubic lattice and those in a face-centered cubic lattice. Compare both to dislocations in a body-centered cubic lattice.

4.9 Show how two edge dislocations of opposite sign on the same slip plane can annihilate each other. Can two screw dislocations of opposite sign also annihilate each other?

4.10 Sketch the distortion of the lattice around an edge dislocation and show the preferred regions for large substitutional atoms, small substitutional atoms and interstitial atoms.

4.11 Low-angle tilt boundaries have an angle of misorientation less than five degrees. Calculate the dislocation spacing in a symmetric 2° tilt boundary in a copper crystal. The edge length of the unit cell is 3.61 Å. What are the Miller indices of the plane of the boundary?

4.12 Face-centered cubic metals usually slip on the most closely packed planes {111} in the most closely packed directions ⟨110⟩. Sketch the (111) plane in a cubic unit cell; draw and index the ⟨110⟩ directions which lie in the plane.

4.13 Calculate the indices of the edge dislocation line which has $\mathbf{b} = \frac{1}{2}[0\bar{1}1]$ and lies in the (111) plane.

4.14 *The indices of a slip plane together with the indices of a slip direction in that plane comprise a slip system.* Possible slip system families in BCC iron are {110} ⟨111⟩, {112} ⟨111⟩ and {123} ⟨111⟩. Sketch a [111] slip direction in each of three cubic unit cells; draw and index three slip planes, one from each of the families above, that produce a slip system when combined with ⟨111⟩.

4.15 Although MgO (NaCl crystal structure) is FCC, slip in it occurs primarily on {110} ⟨110⟩ rather than on {111} ⟨110⟩ as in a FCC metal. Sketch a {110} plane and a ⟨110⟩ direction lying in it.

4.16 Cadmium slips on {0002} in $\langle 11\bar{2}0\rangle$ directions; how many slip systems does it have? Titanium slips on $\{10\bar{1}\bar{1}\}$ in $\langle 11\bar{2}0\rangle$ directions; how many slip systems does it have?

4.17 Hexagonal crystals may twin on $\{10\bar{1}2\}$ in a $\langle 10\bar{1}1\rangle$ direction. Sketch a cross-section of such a twin in a hexagonal Bravais lattice.

4.18 Grain boundaries and twin boundaries are sometimes called "high-angle boundaries." Can they be described in terms of dislocation arrays? Explain.

CHAPTER FIVE

Noncrystalline Solids

Many solids are not crystalline. Their structures are not composed of repetitive, three-dimensional patterns of atoms, nor can they be described simply in terms of imperfections in ideal crystals. However, even though they are noncrystalline, these solids exhibit some local order, either in the form of regular coordination polyhedra or of long chain molecules; they lack the long-range order of crystals because their subunits are packed together randomly. Sometimes these solids are referred to collectively as glasses. They include a few of the elements and many polymers of high molecular weight as well as the common oxide glasses and a few other inorganic compounds. Their atoms rearrange themselves into hard noncrystalline configurations because they have only limited mobility at the equilibrium solidification temperature and normally cannot move into lowest energy (crystalline) configurations.

5.1 INTRODUCTION

At low temperatures, a noncrystalline structure does not have as low energy as a crystal of the same composition; yet, in some materials noncrystalline structures are not only easily formed but are, for all practical purposes, stable. The structures of these materials and the reasons for their stability are important to a general understanding of engineering materials.

The structures of noncrystalline solids are not completely random, for the atoms within the subunits (see Chapter 2) of most of them are ordered. The word "amorphous" was used to designate such structures before X-ray diffraction showed that they had short-range order, that is, order on a scale of the subunit size.

For example, "amorphous carbon" was named before X-ray diffraction showed that its structure was simply one in which individual graphitelike sheets (see Figure 3.9) were parallel but misoriented with respect to one another, that is, it had two-dimensional, but not three-dimensional, order. An X-ray pattern of crystalline powder (see Appendix IV) consists of many sharp lines. In contrast, an X-ray pattern of a noncrystalline powder shows only two or three very diffuse bumps, indicating that the structure has some short-range order but no long-range order. In other words, the atoms are packed within the subunits in an orderly way rather than in a random arrangement as they would be in a truly amorphous solid; the lack of long-range order in noncrystalline solids is due to the fact that these subunits are not packed together in any regular pattern.

Structurally, noncrystalline materials may be classified according to whether they are composed of individual long chain molecules, three-dimensional networks or arrangements somewhere between these two limiting cases. Chemically they may be either elements or compounds. Despite their differences, though, noncrystalline materials have much in common with each other. They do not represent as low energies as do crystalline arrangements of the same atoms and molecules. They respond similarly to changes in temperature in that they do not have a definite melting point but soften gradually as the temperature is raised and harden gradually as the temperature is lowered. When measured at comparable temperatures, relative to their bonding energies, they often have similar physical and mechanical properties.

5.2 GENERAL FEATURES OF NONCRYSTALLINE STRUCTURES

All noncrystalline materials have one feature in common: their structures are such that the subunit arrangements can get tangled so easily and so completely in the liquid state that they are almost impossible to untangle once the material is solid. The ways in which subunit arrangements become tangled are many, but they all may be discussed in terms of two limiting cases: three-dimensional networks and long chain molecules. Three-dimensional networks occur when the subunits are coordination polyhedra which share only corners by means of bonds which are discrete (have a fair amount of covalent character) but flexible. This type of bonding permits many arrangements of the coordination polyhedra which are almost equivalent in terms of energy because the amount of orbital overlap differs very slightly from one to another. Since only a few of these arrangements can be crystalline, most of them will represent noncrystalline structures. The other limiting case of a noncrystalline solid occurs when the subunits are very long, flexible molecular chains which become tangled with each other in much the same way as mile-long pieces of spaghetti might become tangled during cooking. In this analogy, a crystal corresponds to all the pieces of cooked spaghetti straightened out and lined up in a regular array and has a low probability of occurring spontaneously. In both of these limiting cases, as well as those falling between the two, the subunits have too little mobility in the liquid at the equilibrium freezing point. In other words, once the material is cold, the structure remains noncrystalline indefinitely because more atomic motion than is possible would be required to crystallize it.

The similar response which all noncrystalline materials have to changing temperature is related directly to their similarities in structure and bonding. Noncrystalline materials do not have a solidification temperature, as crystalline materials do; they gradually become more viscous over a range of temperature. This may be considered to be a solidification process associated with a range of energies for the bonds between subunits. Since all subunits do not have identical surroundings, they do not have identical bond energies, even though the differences may be small. As a material with this type of structure cools, the lowest energy (most negative) bonds form first and begin to "stick" the subunits together locally; then, as the temperature is lowered further, the weaker bonds gradually form until the material is completely hard.

The temperature at which this solidifying material first seems to become a rigid mass is called the *glass transition temperature, T_g*, because at much lower temperatures the material will be "glass-brittle," and at much higher temperatures it tends to flow like a very viscous liquid.

Besides their common response to temperature, many noncrystalline materials are transparent, both in the liquid and solid states. Their transparency arises because they have no inclusions, holes, or internal surfaces with the right properties to scatter light, and they have no free electrons or ions which can absorb and emit light by changing their energy states. In addition to some similarities in physical properties like transparency, many noncrystalline materials also have similar mechanical properties, as will be discussed in Volume III, *Mechanical Behavior.*

An understanding of the general features of noncrystalline structures provides only a background for understanding individual materials. However, it is a useful background, as will be amplified in the next sections of this chapter.

5.3 ELEMENTS

Although a number of elements with discrete bonds and relatively open structures can be quenched to form glasses at temperatures near absolute zero, sulfur and selenium are the only two in which a noncrystalline structure can be formed by quenching a viscous melt to room temperature. Both elements are in Column VI of the Periodic Table (Appendix IA), and the bonding in both is primarily covalent, due to overlapping *p* orbitals (Chapter 2). This bond arrangement leads to long molecular chains of these atoms, similar to those of tellurium in Figure 3.6. In sulfur and selenium, these molecular chains become so tangled in the liquid that a noncrystalline structure develops when the material is cooled quickly. In so-called fibrous sulfur, the long chain sulfur molecules are actually mixed with S_8 ring molecules, and the one type of molecule keeps the other from crystallizing. The elements, tellurium and polonium, farther down in the same column, do not form noncrystalline solids at room temperature, presumably because the bonding is less discrete and less directional (more metallic, since the electrons are less tightly held), and the atoms and molecules are more mobile. In general, the more diffuse (metallic) the bonding and the more close-packed the crystal structure of an element, the more difficult it is to make it form a glass.

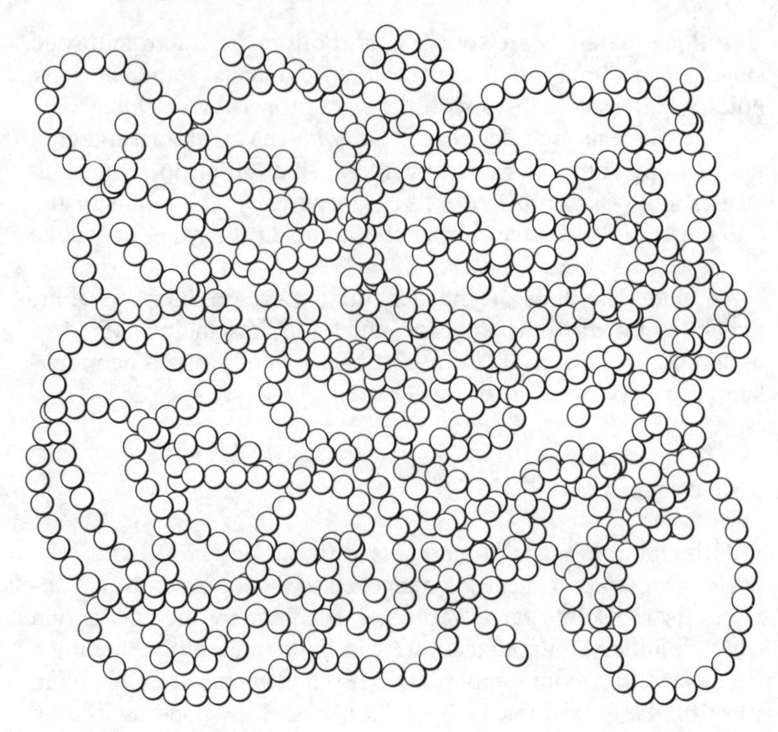

Figure 5.1 A schematic representation of a polymer. The spheres represent the repeating units of the polymer chain, not specific atoms.

5.4 LONG CHAIN MOLECULAR COMPOUNDS

Compounds in which the subunits are long chain molecules are called *polymers*[1] (or occasionally *resins* or *plastics*). They can develop noncrystalline structures relatively easily because their molecular chains are very long and flexible, and can become tangled so easily (Figure 5.1). Polymers have rather loosely packed structures in the solid state because the primary bonds are completely satisfied within the long molecular chains and because the chains often contain bulky side groups of atoms which interfere with close-packing. Any atomic arrangements that contribute to

[1] The structural formulae for a number of typical polymers are given in Appendix V.

the loose-packing of molecular chains will also favor the formation of noncrystalline structures. Among the most important features that favor noncrystallinity are

1. molecular chains that are very long and branched.
2. random arrangements of large side groups along the chains.
3. *copolymer* chains, that is, molecular chains which are actually combinations of two or more polymers.
4. lower molecular weight additives, called *plasticizers,* which separate the chains from one another.

For example, the paraffins (long chain hydrocarbons with the general formula C_nH_{2n+2} and molecular weights up to about a thousand) crystallize almost completely. One form of polyethylene may be regarded as composed of extremely long paraffin chains with molecular weights from ten thousand to a few million. In this form it is called *linear polyethylene* and crystallizes almost, but not quite, as completely as the shorter chain paraffins. In contrast, *branched polyethylene* has side chain segments which are attached to the main chain at positions where a hydrogen atom normally would be found; in this form the polymer crystallizes only partially. The more branching, the more noncrystalline the polyethylene, because the branches interfere with the regular arrangement of chains. Other polymers show similar behavior.

The effect of side group arrangements can be seen by considering the structure of *vinyl polymers,* those polymers with a repeating unit

```
   H  H
   |  |
 —C—C—
   |  |
   H  X
```

where X is some monovalent side group. There are three possible arrangements of side groups (Figure 5.2) in vinyl polymers:

1. *atactic,* or random
2. *isotactic,* or all on the same side of the chain
3. *syndiotactic,* or regularly alternating, from one side to the other.

If the side group is small, as it is in polyvinyl alcohol (X = OH),

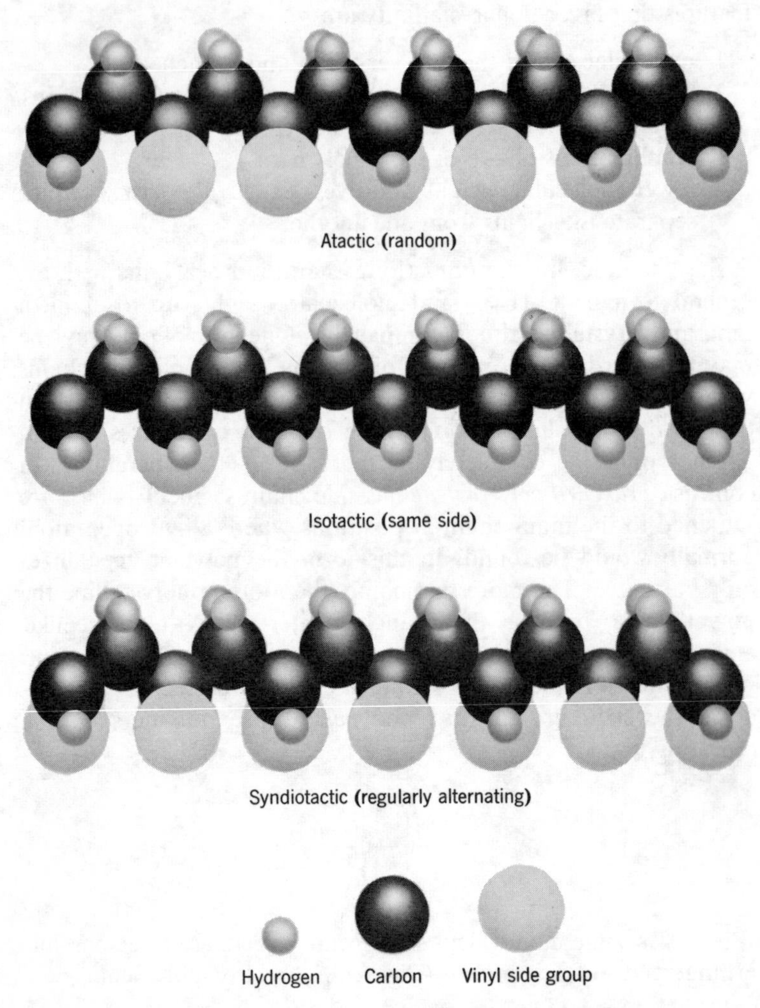

Figure 5.2 The possible arrangements of side groups in a simple vinyl polymer: atactic (random), isotactic (all on the same side), syndiotactic (regularly alternating from side to side). The side group may be a single atom, such as chlorine in polyvinyl chloride; it may also be a group of atoms, such as a benzene ring in polystyrene.

and the chains are linear, the polymer crystallizes quite easily. However, a noncrystalline structure invariably results if the side group is large, as it is in polyvinyl chloride (X = Cl) or polystyrene (X = benzene ring), and is randomly distributed along the chains (atactic). In contrast, isotactic and syndiotactic polymers usually crystallize, even when the side groups are large.

Copolymerization always decreases the regularity of polymer chains and therefore promotes the formation of noncrystalline structures. Copolymers can be made in a number of different geometric arrangements, the simplest of which are illustrated in Figure 5.3. The more irregular, or random, the arrangements, the greater the tendency for noncrystallinity. Quite often a copolymer is developed because a certain amount of noncrystallinity results in better properties. For instance, polyvinylidene chloride, which is normally crystalline and not very pliable, can be made noncrystalline and much more pliable by copolymerizing it with a small amount of polyvinyl chloride. The resulting copolymer is the basis for the Saran plastics.

The addition of plasticizers to prevent crystallization by keeping the chains separated from one another is one of the oldest ways to produce a noncrystalline solid from a polymer that normally crystallizes. One of the first synthetic polymers, celluloid, was made of nitrocellulose (which normally is crystalline) plasticized with camphor. Another common plastic, cellophane, is composed of cellulose chains, prevented from crystallizing by the addition of glycerol as a plasticizer. The disadvantage of this process is that plasticizers are usually of such low molecular weights that they diffuse through the solid and eventually evaporate; the result is a loss of pliability and an increasing tendency to crack with time.

5.5 ELASTOMERS

Elastomers are polymers which exhibit a large and reversible extensibility at room temperature; they can be stretched at least a hundred per cent and often a thousand per cent, and will snap back to their original dimensions once the load is released. Structurally they are all noncrystalline polymers at room temperature and are intermediate between long chain molecules and three-dimensional

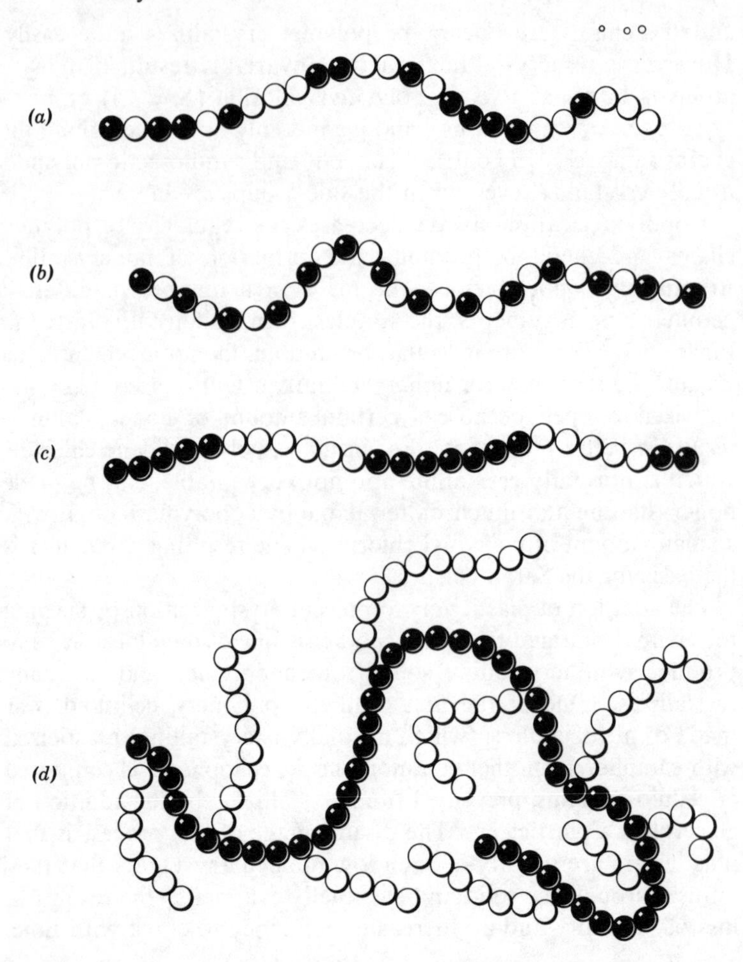

Figure 5.3 Copolymer arrangements: (*a*) A copolymer in which the two different units are distributed randomly along the chain. (*b*) A copolymer in which the units alternate regularly. (*c*) A *block* copolymer. (*d*) A *graft* copolymer.

networks. The structural criteria which they must satisfy, beyond being noncrystalline, are the following.

1. The chains must be very long, with many "bends" in them (i.e., not straight).

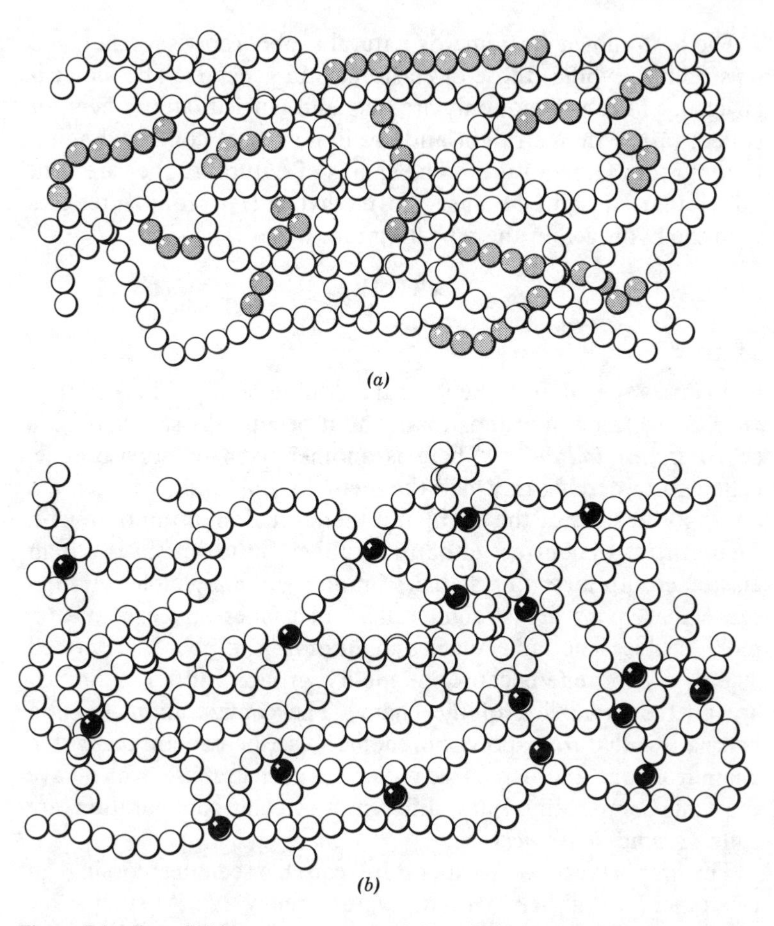

Figure 5.4 Cross-linking arrangements: (*a*) Cross-linked polymer chains in which small chain segments act as cross-links. (*b*) Cross-linked polymer chains in which foreign atoms or molecules are the cross-links. An example of this arrangement is vulcanized rubber, in which sulfur atoms are the cross-links between *cis*-polyisoprene chains.

2. Room temperature must provide enough thermal energy for chain segments to be in a state of constant motion.
3. The chains must be connected to one another every few hundred atoms by *cross-links,* atoms or groups of atoms that form primary bonds between chains (Figure 5.4).

The best-known elastomer is natural rubber (polyisoprene), and its structure typifies those of elastomers in general. The molecular chains in rubber not only are long and tangled but are bent, or coiled, rather than straight and are in a state of continual agitation at room temperature. The chains of natural rubber are bent rather than straight because the methyl (CH_3) group interferes with the hydrogen in the repeating unit

$$\begin{array}{cc} CH_3 & H \\ | & | \\ -C & = C- \end{array}$$

and causes the chain to bend at the double bond. This interference of two atomic groups that cannot occupy the same space is called *steric hindrance.* There is another form of polyisoprene, called gutta percha, in which the methyl group and hydrogen are on opposite sides of the chain and therefore do not interfere with each other. In contrast to natural rubber, gutta percha is not an elastomer. In fact, it crystallizes into a rigid, nonpliable solid because the chains are straighter than in rubber and can line up more easily. The molecular configuration in rubber is described as *cis*-polyisoprene because the methyl groups and the hydrogen are on the *same* side of the chain. The configuration in gutta percha is called *trans*-polyisoprene because the two are *across* the chain from one another. Two molecules such as these, which have the same composition but a different molecular configuration, are called *geometric isomers.*

The importance of temperature can be seen by cooling an elastomer like rubber to a temperature below T_g. At such a low temperature the polymer is glass-brittle. Since the term, elastomer, is restricted by convention to room temperature, this means that the glass transition temperature of a polymer must be well below room temperature for it to be an elastomer. Temperature is so important because the chains in any polymer must be very flexible in order for it to be stretched a few hundred per cent.

The third criterion, cross-linking, is necessary to explain the reversibility of an elastomer's extension. If the chains were not cross-linked, the elastomer would not return to its original shape and size after the force was released. The cross-links act as pinning points; without them, the polymer would deform permanently. Again, the most common example is natural rubber, which simply flows like an extremely viscous liquid at room temperature if not cross-linked. A few per cent sulfur is added to rubber (in a process called vulcanization) to cross-link the polymer by breaking the double C=C bonds and forming C—S—C bonds between the chains. A polymer, like polyethylene, that has no double bonds to break is more difficult to cross-link. However, cross-linking single-bond polymers can be accomplished by subjecting them to radiation of suitable energy, for example, by irradiation in a nuclear reactor. Polyethylene can be cross-linked by irradiation and, in this irradiated form, has a lower tendency to crystallize and, rather than melting as the temperature is raised, it enters into a rubbery state. The cross-links in elastomers are typically a few hundred carbon atoms apart. This feature of their structure actually places elastomers intermediate between the two limiting cases of independent long chain molecules and three-dimensional networks. As the amount of cross-linking increases, the structure becomes more rigid, like a network polymer.

5.6 THREE-DIMENSIONAL NETWORKS

A few polymers form rigid three-dimensional networks. Perhaps the most common network polymer is phenol-formaldehyde (better known by its trade name, Bakelite) in which cross-links are formed by means of the phenol rings which are integral parts of each chain. Such a network polymer does not really have a glass transition temperature; it degrades (depolymerizes) at high temperatures. Another example of a network polymer is ebonite, which is the name given to rubber with a very high density of sulfur cross-links. In this form, the sulfonated rubber is no longer an elastomer but a rather rigid solid because the network is so complete.

Most noncrystalline, three-dimensional networks, however, are inorganic glasses, especially oxides. As with long chain polymers, any atomic arrangements that contribute to the loose-packing of the network constituents will also favor the formation of noncrystalline structures. Of course, the reasons for open structures in inorganic compounds like oxides are different from those that apply to polymers because the subunits are so different. In the oxides, the open structures are usually due to a strong cation-cation repulsion that is most pronounced when the coordination polyhedra are small and the charge on the cation is high. An inorganic compound tends to be noncrystalline if

1. each anion is bonded to only two cations.
2. no more than four anions surround a cation.
3. the anion polyhedra share corners, but not edges or faces.
4. the compound has a large number of constituents distributed irregularly through its network.

For example, silica (Si^{+4} tetrahedrally coordinated by oxygen) and boron oxide (B^{+3} triangularly coordinated by oxygen) form glasses easily. In both structures the cation charge is high and the anion polyhedra are small, leading to very open structures. On the other hand, magnesium oxide (Mg^{+2} octahedrally coordinated by oxygen) does not form a glass; the coordination polyhedra share edges in a relatively close-packed crystal structure.

Most inorganic glasses are based on silica, SiO_2, and therefore have as their fundamental subunit the Si—O tetrahedron (see Chapter 2). In pure silica glass these tetrahedra are joined corner-to-corner but have no long-range order as they do in quartz or cristobalite (see Chapter 3). A schematic representation of what the local structure might be in a silica glass is shown in Figure 5.5, both in terms of silicon and oxygen atoms and in terms of silicate tetrahedra. In B_2O_3, the units are triangles (actually, very flat pyramids, since the boron atom is slightly out of the plane of the oxygen atoms) which are also bonded corner-to-corner in a random configuration. Noncrystalline BeF_2, GeS_2, and As_2S_3 also have their subunits arranged randomly.

Since it is the open, three-dimensional network in many inorganic oxides that is responsible for their noncrystalline structures, any additions that break up the network and facilitate a more close-packed arrangement should probably promote crystallization. This is the case. For example, a silica network can be broken up by adding alkali oxides like Na_2O or K_2O. The oxygen atoms from these oxides interrupt the network by entering it at points where two tetrahedra are joined and breaking them apart so that each tetrahedron has one free corner (one unshared oxygen). The

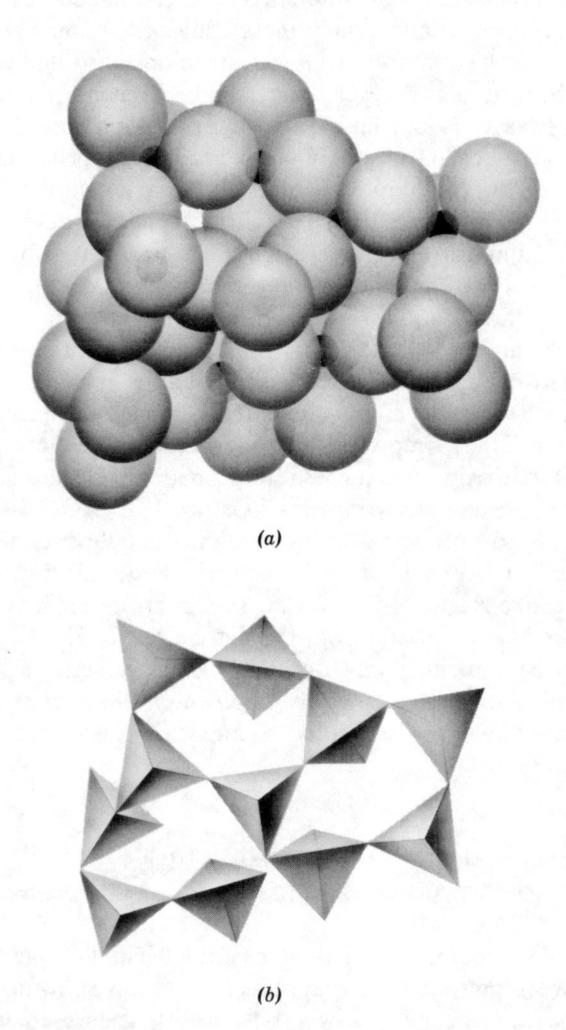

Figure 5.5 (*a*) A schematic representation of silica glass in which oxygen atoms are represented by large spheres and silicon atoms by small dark spheres. (*b*) The identical silica glass structure to that in (*a*), but represented in terms of the arrangement of Si—O tetrahedra. The corners of the tetrahedra correspond to the centers of the oxygen atoms in (*a*).

alkali metal ions enter the interstices of the structure, resulting in closer packing. Commercially these additions are made to break up the network sufficiently to lower the viscosity so that the glass can be fabricated at a lower temperature. General practice is to add both Na_2O (soda) and CaO (lime); this produces common window glass, called a soda-lime-silica glass. If such additions raise the O/Si ratio above about $2\frac{1}{2}$ or 3, the network may be so interrupted that it crystallizes ("devitrifies", in glass terminology). Crystallization can be prevented in marginal cases by adding other glass-forming oxides to the network to make the subunits more irregular. It should be emphasized that both the metal and the oxygen atoms of these glass-forming oxides must become part of the network in order to decrease the tendency toward crystallization. If the metal atoms are added as ions in the interstices of the network, they simply make the structure more close-packed.

B_2O_3 is one oxide which goes into the network of silica glass and may help prevent devitrification. Oxides like B_2O_3, SiO_2, and GeO_2 that contribute to the formation of a three-dimensional network, and form glasses easily themselves, are called *glass formers;* those like Na_2O, K_2O and CaO that interrupt the network are called *modifiers.* In addition, there are a few oxides like Al_2O_3 that may behave either way, depending on such factors as over-all composition; these are called *intermediates.* Either glass formers or intermediates may be added to an oxide glass to prevent its crystallization.

5.7 REPRESENTING THE DEGREE OF ORDER IN NONCRYSTALLINE SOLIDS

Since the structure of a glass does not have long-range order, it cannot be represented by diagrams or pictures showing precise atom locations in space, as crystals are often represented. The structure can be represented by a probability distribution curve that shows the probability of finding another atom at a certain distance from a given one. Such distribution curves, two of which are shown in Figure 5.6, are called *radial density plots.* They indicate how closely a glass approaches being completely amorphous

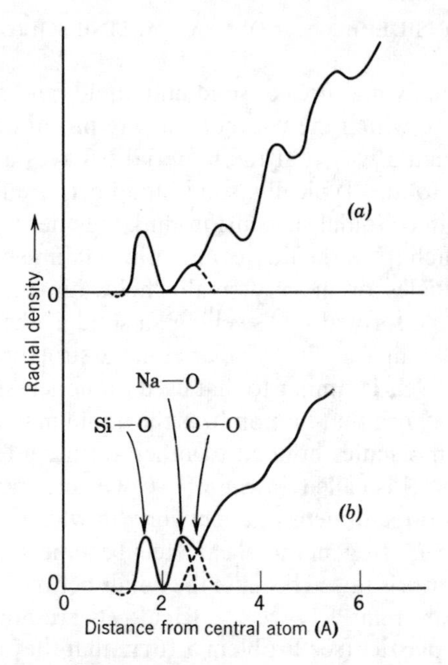

Figure 5.6 Radial distribution curves for (*a*) pure silica glass and (*b*) soda-silica glass, containing 35% soda (Na_2O). (Figure courtesy of G. O. Jones, *Glass*, Methuen & Co., London, 1956.)

or completely crystalline. A completely amorphous material would have a parabolic radial density plot, for the probability of finding another atom at some distance from the first one would be dependent only on the volume available at that distance. All atoms at a distance r from the origin must have their centers on the surface of a sphere of radius r, and the number that can be accommodated is proportional approximately to $4\pi r^2$. At the other extreme, the radial density plot of a perfect crystal would have infinitely high lines at possible interatomic distances. Therefore any bump above the parabola measures the more-than-random probability of finding another atom within the range of interatomic distances given by the width of the bump. The area under the bump is a measure of the number of atoms within this range of distances.

5.8 GELS CONSIDERED AS NONCRYSTALLINE SOLIDS

Gels are actually mixtures of solid and liquid (and sometimes of solid and gas) in which the components are mixed on such a fine scale and in such a way that the material behaves as if it were a noncrystalline solid. Typically, solid-liquid gels are formed when solid particles in colloidal suspension[2] link together to form a solid framework which traps the liquid, either in very fine capillaries between the particles or in very small "holes" in the framework. Gels may also be formed by "swelling" a solid, if the liquid penetrates the solid along very fine capillaries in the structure. The usual classification of gels is similar to that used for noncrystalline solids. The solid framework that immobilizes the liquid may be composed of long chain molecules bonded together at only a few points, in which case the gel is called an *elastic gel.* Or, the framework may be more like a three-dimensional network, in which case the gel is called a *rigid gel.* In general, elastic gels become more liquidlike as the temperature is raised because the bonds between the particles, or molecules, are relatively weak. Rigid gels are much more like thermosetting plastics (see Problem 5.16) in that they do not soften appreciably as the temperature is raised.

Perhaps the best-known example of an elastic gel is not even an engineering material; it is gelatine (one form of which is familiar as "Jello"). In gelatine, the solid framework is composed of long chain protein molecules which trap water between them. Another example of an elastic gel is cellulose gel, which occurs during the making of cellophane. One step in the manufacture of cellophane involves precipitating cellulose from a water solution of one of its compounds; the resulting gel is a three-dimensional arrangement of cellulose chains in which one may picture crystalline regions acting as cross-links and water acting as a plasticizer, separating the chains from one another. In the subsequent operations of the manufacturing process, this elastic cellulose gel is dried; then glycerol is substituted for water as a plasticizer, producing a tougher and less elastic film of cellophane.

Probably the most complex elastic gel of engineering importance

[2] Colloidal particles may be defined as having average dimensions less than 1000 Å. A colloidal suspension is composed of many such particles dispersed uniformly in a liquid with no tendency to settle or coagulate.

is asphalt, a combination of high molecular weight hydrocarbons in an oily residue. The high molecular weight hydrocarbons which make up the solid framework of the gel are called *asphaltenes,* a name which substitutes for a certain amount of ignorance, for their exact chemical constitution is not known. Asphalt is too viscous to be called a true gel at room temperature; however, it gels at lower temperatures. Also, it can be made more gel-like at room temperature by blowing air through it above about 200°C. The air apparently oxidizes the oils and some of the other residues, resulting in a *blown asphalt* that has a higher softening temperature, greater elasticity and a structure that has a more definite framework.

Silica gel is one of the simplest examples of a rigid gel. It is produced by polymerizing monosilicic acid molecules ($Si(OH)_4$) into a rigid, three-dimensional silica network, the interstices of which contain water molecules that are by-products of the polymerization process. Another rigid gel, *Portland cement gel,* is an important engineering material; it is produced during the setting of cement. The finely ground, dry cement contains four principal compounds:

tricalcium silicate	$3CaO \cdot SiO_2$*
dicalcium silicate	$2CaO \cdot SiO_2$*
tricalcium aluminate	$3CaO \cdot Al_2O_3$*
tetracalcium aluminoferrite	$4CaO \cdot Al_2O_3 \cdot Fe_2O_3$*

When this cement is mixed with water, the hydration, or adsorption of water, produces a gel composed principally of hydrated calcium silicates with formulae that are represented approximately by

$$Ca_3Si_2O_7 \cdot 3H_2O$$

Since this is close to the formula for the mineral, tobermorite, the gel is called *tobermorite gel.* The silicate part of the gel is similar to a distorted sheet silicate with Ca and O ions sandwiched between two rumpled silicate sheets. The water molecules are adsorbed on the surfaces of these sandwiches and separate them in the gel. Once this gel is formed, it binds the other particles in the aggregate into a stiff mass which "sets." The mass then hardens further as hydration progresses into the core of each particle.

* These formulae do not represent the structures of the compounds, only the principal starting materials.

DEFINITIONS

Atactic: Having side groups distributed randomly along a vinyl polymer chain.

Branched Polymer: A polymer having secondary chains branching from the main molecular chains.

Copolymer: The combination of two polymers into the same chain; the resulting configuration of the two may be random, alternating, block or graft.

Cross-Linking: The bonding of polymer chains together at various points by means of primary bonds.

Elastomer: A noncrystalline polymer which can be stretched to more than twice its original length and which contracts quickly on releasing the load.

Gel: A solid framework of colloidal particles linked together and containing a fluid in its interstices; depending on the nature of the linking and the geometry of the interstices, the gel may be either rigid or elastic.

Glass: Any noncrystalline solid; applied more commonly to noncrystalline inorganic oxides than to noncrystalline polymers.

Glass Former: An oxide which forms a glass easily; also an oxide which contributes to the network of silica glass when added to it.

Glass Transition Temperature: The center of the temperature range in which a noncrystalline solid changes from being glass brittle to being viscous.

Intermediate: An oxide which may act either as a glass former or as a modifier, depending on the composition of the glass.

Isomers: Compounds of the same composition but different molecular or atomic configurations.

Isotactic: Having side groups all on the same side of a vinyl polymer chain.

Linear Molecule: A long chain molecule with no side branches.

Modifier: An oxide which breaks up the silica network when added to silica glass, thereby promoting crystallization.

Network Structure: An atomic or molecular arrangement in which primary bonds form a three-dimensional network.

Plasticizer: A lower molecular weight material added to a polymer to separate the molecular chains and prevent crystallization.

Polymer: A solid composed of long molecular chains; also referred to as a plastic or a resin.

Steric Hindrance: The interference of two or more atomic groups which cannot occupy the same space.

Syndiotactic: Having side groups distributed in a regularly alternating manner along a vinyl polymer chain.

Vinyl Polymer: A polymer in which the repeating unit of each molecule may be written

```
    H  H
    |  |
  —C—C—
    |  |
    H  X
```

where X is the vinyl side group.

BIBLIOGRAPHY

INTRODUCTORY REFERENCES:

C. H. Greene, "Glass," *Scientific American,* Vol. 204 (January 1961), p. 92.

H. F. Mark, "Giant Molecules," *Scientific American,* Vol. 197 (September 1957), p. 81.

SUPPLEMENTARY REFERENCES:

Z. D. Jastrzebski, *Nature and Properties of Engineering Materials,* John Wiley and Sons (1959), Chapter 2.

G. O. Jones, *Glass,* John Wiley and Sons, New York (1956), Chapters I–III.

B. E. Warren, "Summary of Work on Atomic Arrangement in Glass," *J. Am. Ceram. Soc.,* **24** (1941) p. 256.

MORE ADVANCED TEXTS:

F. W. Billmeyer, *Textbook of Polymer Chemistry,* Interscience Publishers, New York (1957).

E. G. Rochow, *An Introduction to the Chemistry of the Silicones,* John Wiley and Sons, New York (1951).

PROBLEMS

5.1 Give reasons for the conditions, stated in the text, which favor the solidification of a long chain polymer in a noncrystalline structure.

5.2 Give reasons for the conditions, stated in the text, which favor the solidification of oxide in a noncrystalline structure.

5.3 Below their glass transition temperatures, most inorganic glasses are brittle and hard; most organic glasses are brittle and soft. Explain this difference in properties by referring to their respective structures.

5.4 Compare the structure of a silicone with that of a single-chain silicate. Why is the silicone more likely to be noncrystalline?

5.5 How would you design the structure of a silicone to make it an elastomer? What advantages might you expect a silicone elastomer to have over natural rubber? Why?

5.6 Show, from the structural formula of phenol formaldehyde (Bakelite), why this polymer cannot be an elastomer.

5.7 Give the structural formulae for rubber and gutta percha. Show why the carbon chains in the *cis* form of polyisoprene are more likely to be bent and tangled.

5.8 If a long chain molecule of linear polyethylene were scaled up to the diameter of a piece of spaghetti, how long would it be? Assume a molecular weight of 100,000.

5.9 Atactic polystyrene can be "oriented" (have its chains aligned) by stretching it above T_g, but it does not crystallize; rubber, on the other hand, both crystallizes and becomes oriented when it is stretched. Rationalize this difference in behavior in terms of the molecular structures.

5.10 Explain why polymethyl methacrylate (Lucite or Plexiglas) is considerably more transparent than polyethylene at room temperature, while both are transparent above about 125°C.

5.11 Show schematically how nuclear radiation might cross-link polyethylene. Compare the resulting structure with that of cross-linked (vulcanized) rubber.

5.12 Explain why camphor ($C_{10}H_{16}O$) which is normally crystalline promotes noncrystallinity when added to nitrocellulose to make celluloid.

5.13 It has been suggested that glass-formers, intermediates and modifiers may be classed according to their metal-oxygen bond energies. Comment on the validity of this suggestion and if you consider it reasonable, state on which end of the scale the glass-formers would be placed and explain why.

5.14 What imperfections are possible in a glass? How might they be classified?

5.15 The length of a polymer chain is commonly expressed by the *degree of polymerization* (D.P.) which is defined as the number of repeating units in a chain. Calculate the molecular weight of polyvinyl chloride with a D.P. of 1000. Calculate the approximate stretched-out chain length of one of the chains, taking the C—C distance to be 1.54 Å.

5.16 *Thermosetting* plastics do not soften appreciably at elevated temperatures; *thermoplastics* soften and flow easily. Describe the differences in structure which might account for this difference in properties. Give two advantages of thermoplastics.

5.17 Distinguish between a glass transition temperature and a melting point.

5.18 How would you determine the glass transition temperature of a noncrystalline solid?

5.19 The total surface area of the particles in a cement has been measured by water adsorption. A simple cement which was almost fully hydrated had a total surface area of 2×10^6 cm^2/gram. Assuming that the particles were round and had a density of about 1 gram/cm^3, calculate the approximate size of the cement particles. If a colloid is defined as having particles of average dimension less than 1000 Å, could this hydrated cement properly be called a colloid?

5.20 Show how the polymerization of monosilicic acid leads to the formation of silica gel.

CHAPTER SIX

The Shapes and Distributions of Phases in Solids

Most solids have significant structural details on a scale considerably larger than atomic or molecular dimensions. These details depend on the shapes of the different crystalline and noncrystalline aggregates of atoms and molecules (called *phases*) and on their distributions in space. For example, individual crystals often have external shapes which are a result of their structure; they also exhibit crystallographic etch pits, slip lines, and cleavage faces. Polycrystals have grain boundaries separating crystals of different orientations. In addition, a solid may contain more than one phase. In both single-phase and multiphase solids, the shape of a phase and its distribution are two of the most important structural features. Both features may be changed either by thermal processing or by deformation, or by a combination of both. This control over the arrangements of phases in a material makes it possible to optimize and to control properties.

6.1 INTRODUCTION

There are many ways in which crystalline and noncrystalline aggregates of atoms and molecules can be arranged in a solid. These arrangements are responsible for many different structural features on scales larger than those of molecular or atomic dimensions. For example, in a homogeneous, crystalline solid, these features may be the external faces of a crystal, or they may be structure-dependent surface details produced by chemical attack or by deformation. In a homogeneous glass, they may be unique subunit arrangements which are not orderly enough to be called crystalline but which are not completely random. Inhomogeneous solids may have additional structural details due to the ways in which regions with different atomic and molecular structures can be distributed throughout the material. These regions of different structure are called *phases,* and the boundaries between them are called *phase boundaries.*

The shapes and distributions of phases are important features of solids. But they are not static features. They may be changed by thermal and mechanical treatments. The alteration of the structures, shapes, and distributions of phases for specific purposes is called *processing.* Some of the techniques and effects of processing will be discussed after the structural features themselves are described.

6.2 SINGLE CRYSTALS

Perhaps the most obvious manifestation of crystalline order on a microscopic or a macroscopic scale is the shape of a single crystal grown in an unconstrained manner. A photomicrograph of a snowflake grown approximately in this way is shown in Figure 6.1. Note that its symmetry is approximately hexagonal as we might expect from the crystal structure of ice (see Figure 3.15). In addition to growing in geometric shapes, many crystals fracture along specific atomic planes by a process called cleavage, yielding relatively flat faces which meet each other at specific angles determined by the crystal structure. The crystal of calcite ($CaCO_3$) shown in Figure 6.2 was cleaved from a larger crystal which had no regular geometric shape.

Another manifestation of crystallinity arises from the specific attack of some chemicals, called *etches,* which corrode local regions of a solid in preference to others. A cleaved {100}[1] face of a LiF (NaCl crystal structure) crystal is shown in Figure 6.3 after being etched. The square, pyramidal pits are called *etch pits;* they illustrate the cubic symmetry of the {100} face, and the fact that they all have the same orientation means that the entire region in the

[1] The use of indices to describe atomic planes is described in Appendix III.

Figure 6.1 Photomicrograph of a snowflake. The sixfold symmetry arises from the hexagonal atomic arrangement in an ice crystal (see Figure 3.15). (Figure courtesy of Buffalo Museum of Science.)

photomicrograph is one crystal. The origin of each etch pit on this crystal face is the distortion around a dislocation (see Chapter 4) where it intersects the surface. Therefore, besides indicating the approximate orientation of the crystal, etch pits also may show how some of the dislocations are distributed.

Other structural details develop on the surfaces of ductile single crystals when they are deformed. As planes of atoms slip over one another, steps develop on a previously polished surface so long

Figure 6.2 A single crystal of calcite, $CaCO_3$, sometimes called iceland spar. The crystal has the shape of a rhombohedron.

Figure 6.3 An etched {100} face of a LiF single crystal. The square pyramidal pits result from local chemical attack around dislocations and are called etch pits. (Figure courtesy of J. J. Gilman.)

as the Burgers vectors of the dislocations producing slip are not parallel to the surface. These steps look like lines in the microscope and are called *slip lines.* Examples are shown in Figure 6.4.

6.3 SINGLE-PHASE POLYCRYSTALS

A solid may be single phase and yet consist of an aggregate of crystals having different orientations. The individual crystals in such a polycrystalline aggregate usually are called *grains:* the surface imperfections separating neighboring grains are called grain boundaries and were discussed in Chapter 4. The grains of a fractured zinc ingot are shown in Figure 6.5, both as they appear on the fracture-surface and as they have been etched on the top surface of the ingot. Usually, however, the grain structure of a material is on a much finer scale. If such a fine-grained material is opaque, it may be examined by polishing a smooth, flat surface on it, then etching it and viewing it in a reflected-light microscope. The etching agent may attack grains of different orientations differently, or it may delineate the structure by attacking the grain boundaries. The etch used for the brass (70% Cu, 30% Zn) sample shown in Figure 6.6 did both. The irregular regions outlined in the photomicrograph are the grains; the parallel-sided regions within each grain are twins (discussed in Chapter 4). Since the etch attacks different orientations differently, a twin in a grain of brass etched this way looks either lighter or darker than the matrix of the grain. If the polycrystalline solid is transparent, the grain boundaries may be observed also by polishing a section about 0.001 inch thick and viewing it in transmitted light. The grain boundaries may be imperfect enough to scatter light and they will then appear as dark lines in the microstructure.

Polycrystalline polymers also have a grain structure, which shows up particularly well in transmitted, polarized light. An example is shown in Figure 6.7. Here, however, the structure within the grains is not quite the same as it is in an inorganic crystal; the orientation of the molecular chains varies gradually from one part of a grain to another rather than remaining constant throughout as in an ideal crystal. These polymer grains are called *spherulites* (sometimes spelled spherolites) because, although the *local* pack-

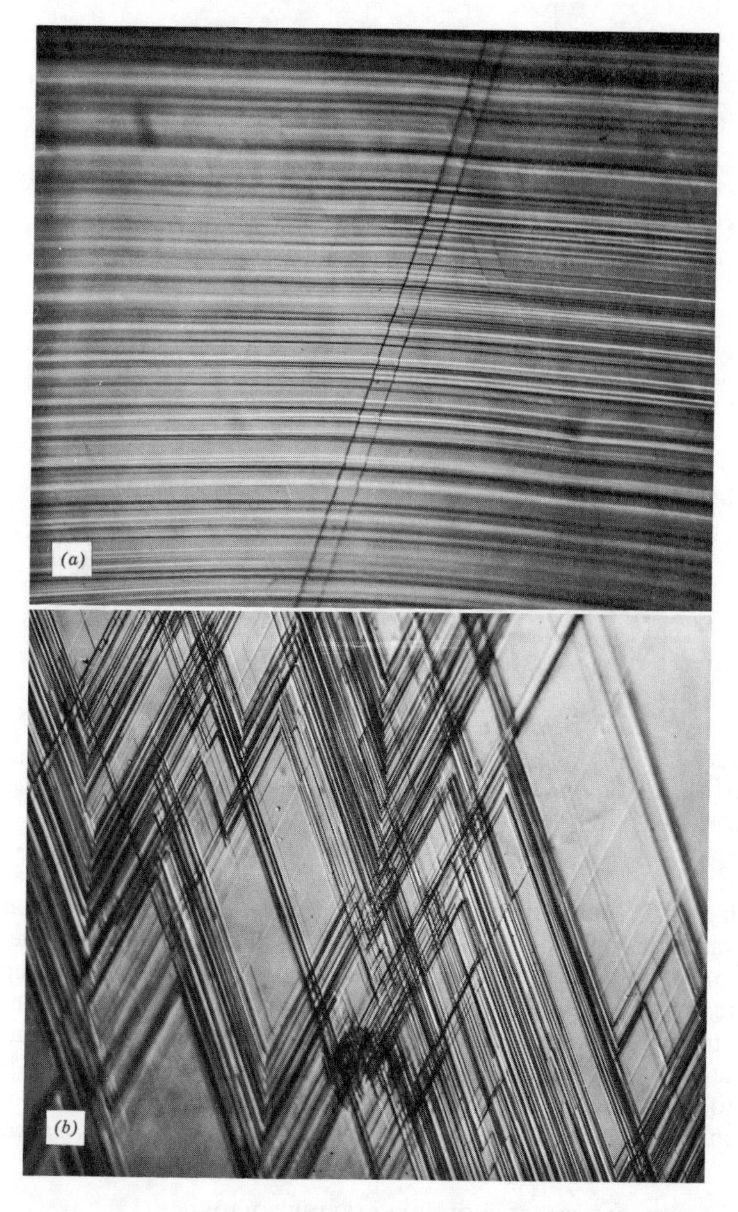

Figure 6.4 Slip lines formed during the deformation of a copper-7 wt.% aluminum single crystal. Intersecting lines result from the operation of more than one slip system of the type $\{111\}\ \langle 1\bar{1}0 \rangle$. (Figures courtesy of G. A. Miller.) (*a*) 400X. (*b*) 200X.

Figure 6.5 The fracture surface of a zinc ingot, showing the facets produced by the cleavage of individual grains (crystals) in the polycrystalline aggregate. The grains also have been delineated on the top surface of the ingot by etching.

ing of the polymer chains is on a space lattice, their orientation varies throughout the grain in such a way that the molecular structure of the grain itself has approximately spherical symmetry. The reasons for this spherical symmetry are not known completely but appear to be related to the way in which molecular chains, thousands of Angstroms long, attach to a crystal nucleus and cause it to grow. Some of the current ideas concerning the structure of spherulites will be presented in Volume III, *Mechanical Behavior.*

6.4 SINGLE-PHASE, NONCRYSTALLINE SOLIDS

In general, noncrystalline solids have no recognizable structure on a scale larger than a few interatomic distances. The reason for this apparent deficiency is simply that our concept of recognizable structure implies some degree of orderliness, and most noncrystalline solids exhibit no order in the way their subunits are packed together. The few exceptions are solids in which the subunits are long molecular chains with weak interchain bonding (e.g., polymers) and solids in which the subunits are sheets weakly bonded to one another (e.g., graphite). These exceptions are described briefly in the following two paragraphs.

Figure 6.6 A reflected light photomicrograph of the microstructure of 70-30 brass (70% Cu, 30% Zn), showing many grains of different orientations and showing twins (the parallel-sided regions) within the grains. 500X.

Figure 6.7 A transmitted light photomicrograph of the grain structure of polyethylene. Each polygonal region is a spherulite. The black "cross" in each spherulite shows the orientation of the crossed polars used to view the polymer. 525X. (Figure courtesy of F. P. Price.)

Order can be developed in some noncrystalline polymers by stretching them above their glass transition temperatures to increase the alignment of the chains. If the polymer is atactic with large side groups, it will not crystallize, but its structure in the stretched condition will definitely be more orderly. A polymer in which most of the molecular chains have been aligned like this is said to be *oriented.* In a few polymers, like *cis*-polyisoprene (natural rubber), this orientation may actually result in crystallization, but atactic vinyl polymers with large side groups, like polystyrene, exhibit no such tendency.

In graphite (see Figure 3.9 for the crystal structure), the carbon atoms in each sheet always have a very orderly arrangement. But the sheets of carbon atoms sometimes are rotated with respect to one another. If they are oriented perfectly, the result is crystalline graphite; if they are not oriented but, rather, are rotated randomly, the result is noncrystalline and is called "amorphous" carbon. A spectrum of structural arrangements lies between these two extremes and includes carbon black, lamp black, etc. All but crystalline graphite are noncrystalline, but each has a different degree or type of order.

6.5 MULTIPHASE SOLIDS

Most solids contain two or more phases. This multiplicity of phases complicates their structures considerably but also endows them with many useful and interesting properties. A number of structural features are important in multiphase materials, specifically, the nature of each phase, its distribution in the microstructure, the amount of each phase, and the size of its domains. A general scheme for classifying materials according to these features has never been agreed on; therefore we shall discuss multiphase solids by describing some of the more important phase arrangements as they are found in actual materials. We shall start with those materials in which the phases are distributed on a rather coarse (macroscopic) scale, and proceed to those in which the distribution is on a microscopic and a submicroscopic scale.

Perhaps the most common multiphase materials are the naturally occurring solids we call rocks. In the terminology of the geologist, a rock is composed of two or more minerals; in more general terminology we say it is composed of two or more phases. Figure 6.8 is a photograph of a rock containing feldspar and quartz. In general, the more slowly the rock cools during its formation, the coarser the scale of the phase distribution. The rock shown in Figure 6.8 apparently formed quite slowly, for the individual phase domains are so large. Most rocks have finer structures and also many more phases.

Ceramic materials like building bricks and concrete are also multiphase solids. They have some rocklike characteristics due to the hard, brittle nature of their phases, but they differ from rocks both in their origin and in the structures of the phases present. The phases in a rock nucleated and grew during the slow formation of the earth's crust and are usually (but not necessarily) crystalline. The matrix phase in building brick, like that of many ceramics, was formed during firing and is a *glassy* silicate. This glass bonds the particles of sand and unreacted clay into a solid mass. The matrix

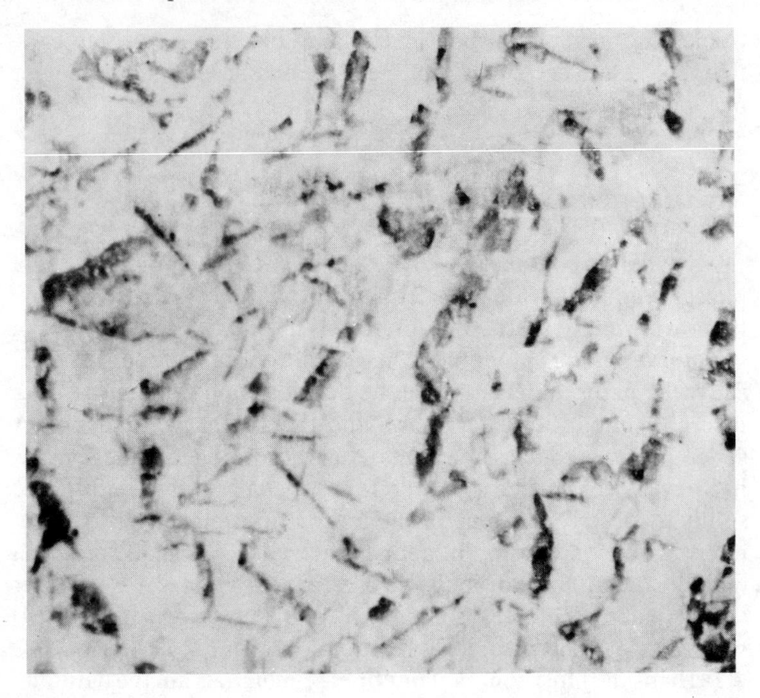

Figure 6.8 The phase distribution in a two-phase rock. The phases are quartz (dark) and feldspar (light). 2X.

phase of concrete, on the other hand, is a silicate *gel*. Concrete is a mixture of sized gravel and sand, bonded together by a hydrated Portland cement gel. This gel was described in Chapter 5 and is really a two-phase mixture on a submicroscopic scale; however, it behaves as a single, noncrystalline phase, and we shall treat it as such. Most multiphase ceramic materials are similar to these two in the sense that the crystalline phases are bonded together by a noncrystalline matrix. Completely crystalline ceramics have been available commercially for a relatively few years. The phases which are present in a ceramic and the scale on which they are distributed depend on how the ceramic is made, that is, on its processing.

Multiphase *alloys* differ from ceramics primarily in the nature of the matrix phase. The matrix phase in an alloy is a metal. The phases in alloys may be pure metals (although this is rare), substitutional solid solutions, interstitial solid solutions or compounds with a fixed chemical composition. Their distributions are much more sensitive to heat treatment than are the phase distributions in ceramics because metal atoms are more mobile in the solid state than ionically and covalently bonded atoms. For this reason it is difficult to characterize phase distributions other than by specific examples.

Steels and cast irons are typical examples of multiphase alloys which are also important engineering materials. *Plain carbon steels* are alloys of iron containing up to about 2 weight per cent carbon; no other alloying elements are added intentionally although impurities are always present. Below 0.8 weight per cent carbon, the matrix phase is *ferrite,* an interstitial solid solution of carbon in iron. When the steel contains between about 0.02 and 0.8 per cent carbon, this ferrite matrix contains lamellae of a second phase, an iron-carbon compound with the formula Fe_3C. The lamellar arrangement of ferrite and Fe_3C is called *pearlite* because it is similar to the lamellar structure of mother-of-pearl, and it is shown at two magnifications in Figure 6.9. When the carbon content is

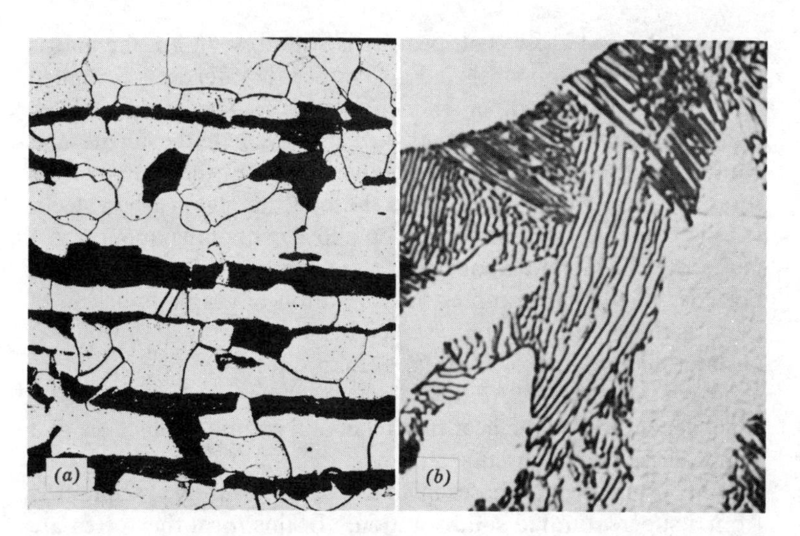

Figure 6.9 Microstructures of a 1020 steel (an iron alloy containing 0.20 wt.% carbon as the major alloying element). (*a*) Pearlite (dark regions) and ferrite in a 1020 steel plate. 126X. (*b*) Same as (*a*) but at 1700X, showing the lamellar structure of the pearlite. The dark lines are phase boundaries between ferrite and Fe_3C in the pearlite.

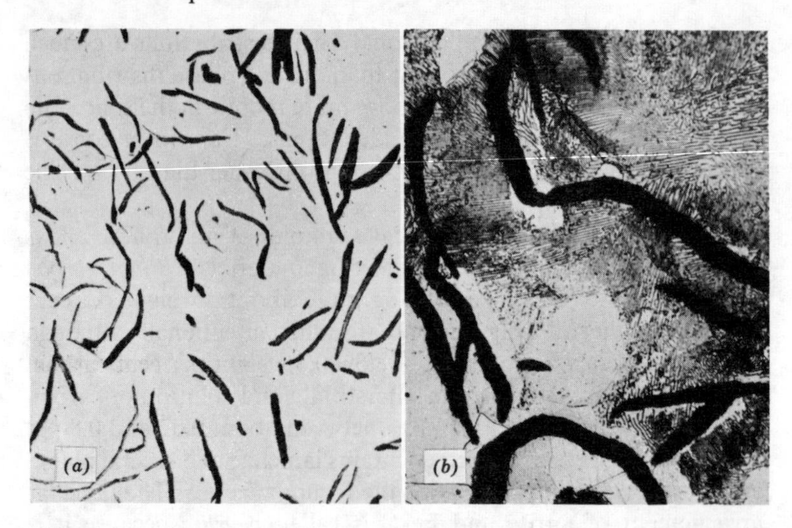

Figure 6.10 Microstructures of a pearlitic gray cast iron. (*a*) Unetched to emphasize the shape and distribution of the graphite flakes. 85X. (*b*) Etched to show the pearlite in the matrix. 425X.

between 0.8 and 2 per cent, pearlite is still present, but the matrix of these high carbon steels is Fe_3C rather than ferrite. So-called alloy steels contain other elements, such as chromium, nickel, manganese, molybdenum, tungsten, dissolved in the ferrite and combined with carbon as complex carbides, but their microstructures are similar in appearance to those of the plain carbon steels.

Cast irons are iron-silicon-carbon alloys containing about 2 to 4 per cent silicon and about the same amount of carbon. The presence of silicon promotes the formation of graphite instead of Fe_3C as the carbon-rich phase when the alloy is cooled slowly. At an intermediate rate of cooling, both graphite and Fe_3C form, as shown in Figure 6.10. Additional phase arrangements are also possible, depending on the heat treatment and composition of the cast iron. These will be discussed in more detail in Chapter 8.

Quite often second phases in an alloy also occur by precipitation from a supersaturated solid solution. In this form they often are too small to be seen in an optical microscope but nevertheless are important sources of strength in alloys. Second phases usually precipitate first at the grain boundaries, where the imperfect nature of atomic packing can accommodate them more easily, but often precipitate later in the grain interiors as well. The lower the temperature at which precipitation occurs, the higher the probability of finding it within the grains and the finer and more general the distribution of precipitate will be.

Wood is another common multiphase solid. However, its structure differs substantially from those discussed above because it originates as part of a living organism. Therefore the microscopic building block of wood is a living cell. The structure and properties of wood are dictated largely by the sizes and distributions of these cells and by the structure of the cell walls. Cells in most trees are very elongated in needle-like shapes and contain a watery solution. The cell walls are composed of bundles of cellulose chains (see Chapter 3 for the crystal structure of cellulose) in which the local molecular arrangement is crystalline. The cellulose bundles spiral around the cell and are bonded in two or three distinct layers into a tough cell wall with a complex noncrystalline polymer called lignin.[2] The lignin also bonds the cells together. Although the cells can be seen easily with an optical microscope, the cellulose-lignin arrangement is on such a fine scale within the cell walls that only rarely can it be resolved, even with an electron microscope. Generally there is about twice as much cellulose as there is lignin. The more lignin, the softer and more "springy" the wood.

Many other multiphase solids in which the matrix phase is a polymer also have fine-scale phase distributions. When the scale approaches molecular dimensions, as when a plasticizer (see Chapter 5) is added to a polymer, it is common to consider the material as "almost" single phase. However, when another phase is added primarily to strengthen a polymer, it is usually present in the form of small particles which are more obviously different from the matrix. These second-phase additions to polymers are called *fillers.* They may comprise as much as 50 per cent of the material's weight, and their distribution within the material is usually random. The filler can be almost anything,[3] but the most effective fillers are those that can form bonds with the polymer

[2] According to Billmeyer, *Textbook of Polymer Chemistry* (Interscience Publishers, 1957) p. 370, lignin is a polymer of various groups, having the carbon skeleton of *n*-propylbenzene.

[3] Common fillers are shredded asbestos, wood flour, cotton fibers, graphite, carbon black, metal powders and fine oxides.

chains. One of the most potent fillers is carbon black, used to strengthen rubber. Apparently strong bonds form between the hydrocarbon chains and active groups on the surfaces of the carbon black particles and behave almost as cross-links.

The fibrous nature of multiphase polymers has suggested a number of synthetic materials in which stronger filaments are substituted for the polymer chains and in which the pores and voids inherent in a cellular material like wood are eliminated. One of these synthetic materials is *fiberglas,* a composite of very fine inorganic glass filaments bonded together with an organic polymer. Depending on the directionality of properties desired, the fibers may be aligned parallel to one another, or they may be wrapped spirally (as if part of a "giant wood cell"), or laid down in random orientations. More recently, investigators also have been experimenting with metal and inorganic whiskers[4] in both polymer and metal matrices. This field of so-called *composite materials,* in which synthetic microstructures are developed for specific purposes, is perhaps the most important new field in engineering materials.

6.6 PROCESSING: MELTING AND SOLIDIFICATION

The fabrication of most metallic and many nonmetallic materials involves melting the raw materials and pouring the resulting liquid into a *mold* which produces a solid of manageable size and shape. Solidification usually proceeds inward from the mold wall, as heat is extracted out through the mold wall. As a result, the grains which form are often *columnar*—long and narrow and perpendicular to the mold wall. The grains usually do not grow homogeneously and instantaneously; rather, each forms a skeletal structure of planes first, the remaining liquid between the planes solidifying later. The skeletal framework of a grain is called a *dendrite* (Figure 6.11) and is similar to the snowflake structure shown in Figure 6.1. Often, dendrites are delineated, as in Figure 6.11*b*, by etching, because their compositions vary from inside to outside. This compositional gradient from the first to the last material to solidify is called *coring* and normally is considered to be undesir-

[4] A whisker is a very fine, strong single crystal filament with either no dislocations or with one screw dislocation running along its axis.

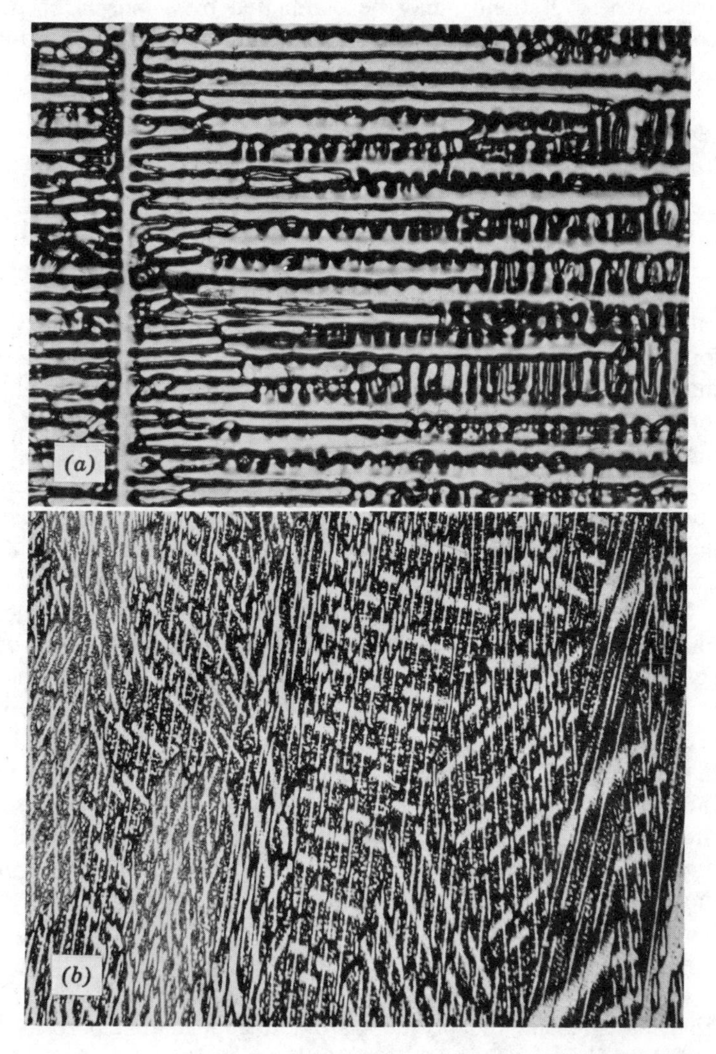

Figure 6.11 (*a*) Dendrites on the surface of a tin casting. They appear as ridges on the surface because the last material to solidify contracts into the spaces left between the dendrite arms which solidifies first. 100X. (*b*) Dendrites delineated by etching an aluminum-4.5 wt.% copper alloy. The cores of the dendrites have a different composition from the last material to solidify and are therefore attacked differently by the etch. 12X. (Both photographs courtesy of T. F. Bower.)

able. These gradients may be eliminated by *homogenizing* the cored casting, that is, by heating it at an elevated temperature until the composition is uniform throughout the material.

6.7 PROCESSING: POWDER PROCESSES

Many solids are not made by melting and casting because it is more economical to form them by mixing their components together as powders or particles, then fusing them together into a solid mass. Ceramic powders are often mixed with water first to form either a slurry or a soft, moldable plastic mass. The shape that the powder-water mixture is poured or pressed into is dried to remove most of the water and then *fired* at an elevated temperature to bond the powders together. As described earlier, the bonding phase is usually a glass. Concretes also are made by mixing powders (dry cement, sand, and gravel) with water, but in this case the cementing phase is produced by a chemical reaction which occurs while the concrete is *setting* at normal temperatures.

Powders may also be bonded together without forming a new phase. If the powders are of inorganic materials, the bonding occurs by solid state diffusion and is called *sintering*. A sintered part is made by pressing the powders together in the desired shape, then heating (sintering) them at temperatures too low to melt the mass but high enough to promote diffusion between particles. If the powders are of thermoplastic polymers, the bonding occurs in the viscous, rather than the solid, state, and the process is called *molding*. Pressure and heat usually are applied at the same time during the molding of these polymers. If the polymers are of the thermosetting type, the powders are normally first pressed into shape, then heated to cross-link them into a solid mass.

In all polymeric materials, the pressures used during molding are high enough, and the powders are soft enough that porosity is generally absent. This is not always so with inorganic materials. Objects made from inorganic compounds by powder processes such as firing, setting, and sintering contain porosity and islandlike microstructures when the powders are too hard to be pressed together so completely that all void space is eliminated. Porosity is a significant phase in these materials; it decreases strength, decreases thermal conductivity, and it can cause an inherently transparent material to appear opaque. Some degree of densification occurs during the bonding process in these materials, but achieving 100 per cent density requires rather elaborate procedures.

6.8 PROCESSING: DEFORMATION

In general, metals and polymers are the only materials ductile enough to be deformed enough that their structures are changed appreciably. The changes produced by deformation are particularly pronounced in metals, and we shall consider them first. In order to make metal sheet, bar and rod stock from previously cast metal ingots, the ingots must be deformed. They may be passed between hard cylindrical rollers in a process called *rolling.* They may be hammered or pressed between dies in a process called *forging.* They may be pushed through a die (like toothpaste is squeezed from a tube) during *extrusion,* or they may be pulled through a die in *wire drawing.* During any of these processes, the grains change shape (see Figure 6.12*a*) as dislocations pass through them. The dislocations passing through the grains on intersecting slip systems interact with each other, producing tangled dislocation arrangements. These tangles hamper the passage of further dislocations and thus make the grains harder to deform. This hardening process is called *strain-hardening* because it occurs when the metal is strained. In addition to becoming harder during deformation, a metal usually develops a *preferred orientation,* that is, the orientations of the grains become aligned with one another. This preferred orientation may be responsible for very directional properties in some materials. Also responsible for directional properties are impurity phases, called *inclusions,* which are drawn out into elongated stringers during deformation. Typically, these inclusions are both hard and brittle at room temperature.

The few nonmetallic crystals, such as NaCl and CaF_2, which are soft enough to be deformed, behave in much the same way as metals: slip occurs in them by the passage of dislocations. However, there are problems associated with deforming polycrystalline aggregates of nonmetallics. These will be discussed in Volume III, *Mechanical Behavior.*

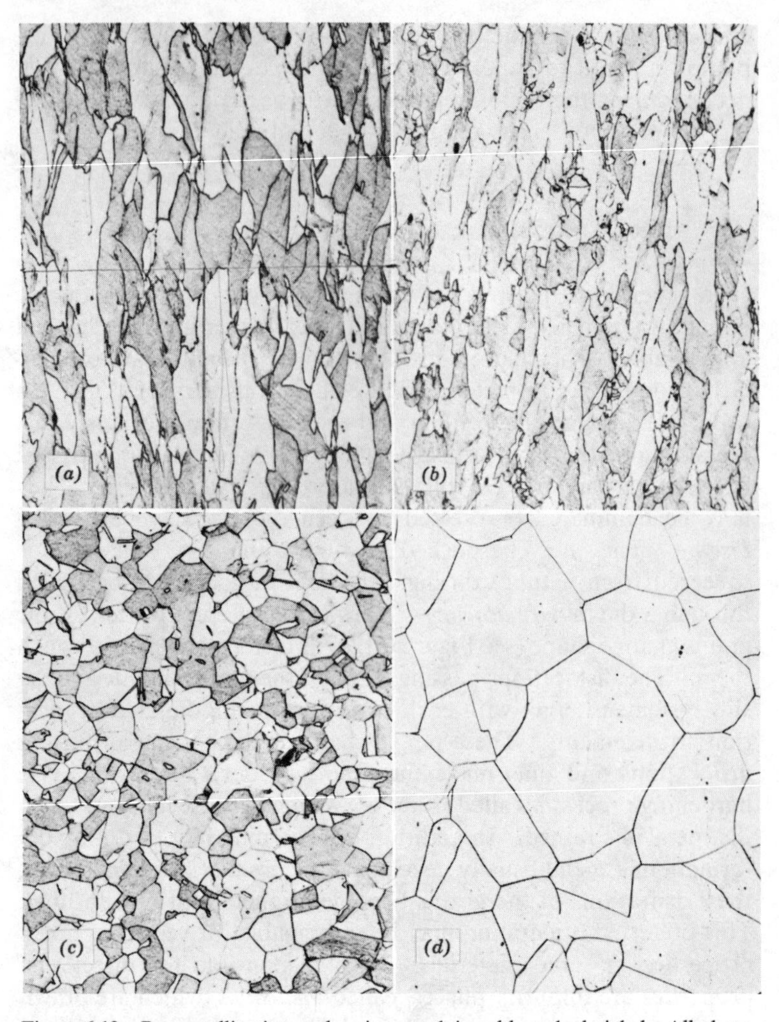

Figure 6.12 Recrystallization and grain growth in cold-worked nickel. All photomicrographs 170X. (*a*) Cold-worked. The direction of working is vertical. (*b*) Partly recrystallized. (*c*) Completely recrystallized. (*d*) After grain growth has taken place. (Figure courtesy of P. A. LaFrance.)

Polymer structures are changed also by deformation, but the change is almost always an increased directionality of the chains—a preferred orientation. As described earlier, polymers may be formed from powder or pellets by molding; this is a process much like forging a metal except that the polymer is being agglomerated into a single mass at the same time it is having its shape changed. Polymer sheets may be formed by *calendering,* an operation similar to rolling except that the starting material is powder or pellets, not an ingot, and agglomeration and shape change take place together. Similarly, *extruding* a polymer usually involves agglomerating it just before it is pushed through a die. One polymer deformation process is particularly unique and has no counterpart in metal-working technology. This is the process of *blowing* a polymer into a "balloon," then slitting the balloon to obtain very thin sheets. These sheets have interesting properties because they are biaxially oriented; the polymer chains are lined up in the plane of the sheet, but their directions in this plane are random.

6.9 PROCESSING: RECRYSTALLIZATION AND GRAIN GROWTH

As a crystalline material is deformed, its hardness increases as the dislocations become more tangled, and its ductility decreases. Typical property changes in a metal are shown in the left-hand part of Figure 6.13. If the material is heated to a slightly elevated temperature, the dislocations of opposite sign begin to annihilate each other; the point defects disappear, and physical properties such as resistivity approach the values typical of the undeformed metal.

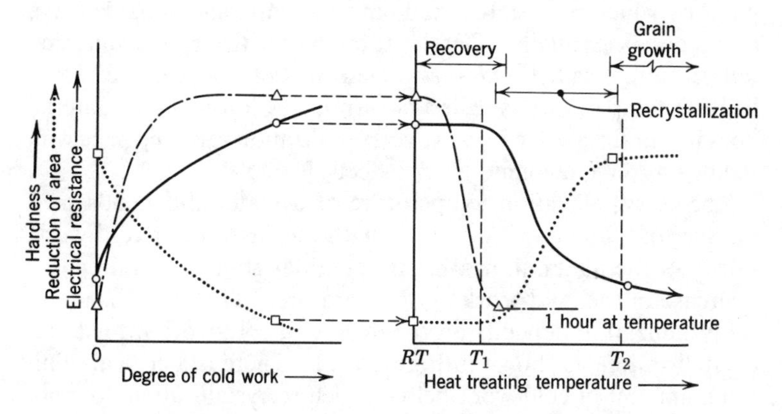

Figure 6.13 Schematic cold-work and annealing cycle.

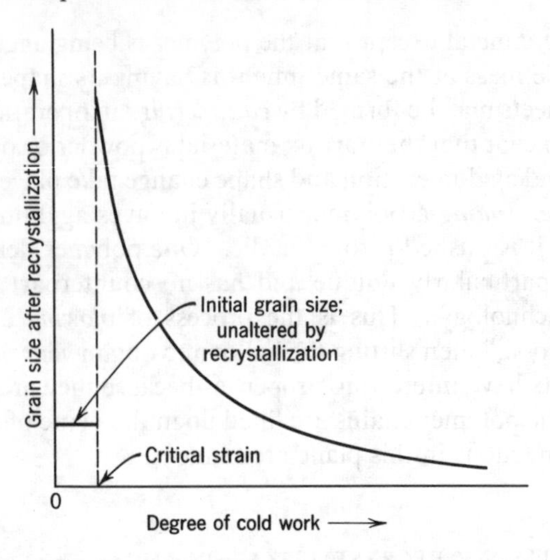

Figure 6.14 Schematic representation of recrystallized grain size as a function of prior cold-work.

This process is called *recovery.* The microstructure, as viewed in the optical microscope, remains unchanged until, at a higher temperature, the elongated grains transform to fine, equiaxed grains, and hardness and ductility approach their original values. This nucleation of new, strain-free grains in a deformed material is called *recrystallization* and proceeds until the entire sample consists of new grains, as in Figure 6.12. The name given to the heat treatment by which recrystallization, and therefore softening, is made to occur is *annealing.* Deformation below the recrystallization temperature is called *cold-work.* Deformation above the recrystallization temperature is called *hot-work.* A hot-worked material does not strain-harden because recrystallization can keep pace with the distortion and counteract its effects.

The recrystallization temperature of a material depends on a number of variables; it decreases with increased prior cold-work, purity of the material, heating time, initial grain size, and with a decrease in the cold-working temperature. The grain size after recrystallization depends on the degree of cold-work prior to recrystallization, as shown in Figure 6.14. There is a critical minimum amount of cold-work below which recrystallization does not occur. If the degree of cold-work is greater than the minimum but still small, relatively few strain-free nuclei form during recrystallization, and the recrystallized grain size is large. The more cold-work, the more nuclei, the smaller the grain size.

Recrystallization requires prior cold-work. *Grain growth* does not. Any fine-grained aggregate of crystals, whether processed by recrystallization, sintering, or some other process, will increase in grain size when heated at an elevated temperature; the "driving force" for this grain growth is the reduction in surface energy by the reduction in grain boundary area. The higher the temperature, the more rapid the grain growth. Figure 6.12*d* shows the microstructure of nickel after some grain growth has taken place. The larger grain size than either the starting material or the recrystallized material is evident.

Other microstructural changes occur during recrystallization and grain growth. We shall describe one that is common among some FCC metals—the development of *annealing twins,* such as the twins in Figure 6.6. Unlike deformation twins, annealing twins do not occur by shear but, rather, as irregularities in the growth of strain-free grains from deformed grains. In FCC metals, the twin boundary is of the family $\{111\}$, and a twin may be visualized as starting when one $\{111\}$ plane of atoms falls into place on the next lower plane in a local HCP configuration rather than continue the FCC configuration. If the stacking of close-packed planes in the perfect FCC matrix is described as ABCABC . . . , the out-of-place plane will change the sequence to ABCABCB . . . . Now if the remaining planes stack in a normal FCC sequence from this plane on, the entire sequence across the boundary will be ABCABCBACBAC . . . ; the structure is "twinned," and the region around the C in the middle of the above sequence represents the twin boundary, as discussed earlier in Chapter 4.

DEFINITIONS

Annealing: Heating a material in order to soften it.

Annealing Twins: Twins produced during recrystallization and grain growth; if produced during crystallization, they may be called growth twins.

Cold-Work: Deformation below the temperature at which recrystallization and extensive recovery take place.

Columnar Grains: Long, thin grains which usually grow preferentially in the direction of the maximum temperature gradient.

Composite Materials: Multiphase materials in which the phase arrangement and distribution is controlled by mechanical rather than thermal and chemical means.

Dendrite: The crystallographic skeleton of a grain, composed of the first solid (usually in the form of crystallographic plates or spines) which crystallizes upon cooling a liquid or a gas.

Etch Pit: A pit produced on the surface of a crystal by chemical attack; it has a geometric shape and is bounded by crystallographic planes.

Filler: A second phase added to a polymer to increase its mechanical strength.

Firing: Heating a ceramic to an elevated temperature to consolidate it and bond it together, usually by forming a glassy matrix phase.

Grain: A single crystal in a polycrystalline aggregate.

Grain Growth: An increase in the average size of the grains in a polycrystalline material.

Homogenizing: Eliminating all concentration gradients in each phase, usually by an elevated temperature heat treatment.

Hot-Work: Deformation above the temperature at which recrystallization and extensive recovery take place.

Inclusion: A foreign or impurity phase in a solid.

Molding: Forming powders of a polymer into a solid mass by the application of pressure and heat.

Phase: A volume of material which contains no discontinuity in either composition or structure.

Preferred Orientation: The preferential alignment of either crystals or molecular chains, producing a similar orientation in every part of the solid.

Processing: The alteration of the structures, shapes and distributions of phases for specific purposes.

Recovery: The first stage in removing the effects of deformation by heat treatment; it precedes recrystallization.

Recrystallization: The nucleation of new, strain-free grains in a deformed crystalline matrix.

Setting: Bonding aggregates of particles together with a cementing phase produced by a chemical reaction at ambient temperatures.

Sintering: Bonding powders together by solid state diffusion, resulting in the absence of a separate bonding phase.

Spherulite: A grain of a crystalline polymer within which the orientations of molecular chains vary so as to give the molecular structure of the grain approximately spherical symmetry.

BIBLIOGRAPHY

INTRODUCTORY REFERENCES:

R. D. Preston, "Cellulose," *Scientific American,* Vol. 197 (September 1957) p. 156.

G. Slayter, "Two-Phase Materials," *Scientific American,* Vol. 206 (January 1962) p. 124.

SUPPLEMENTARY REFERENCES:

R. M. Brick and A. Phillips, *Structure and Properties of Alloys,* McGraw-Hill Book Co., New York (1949).

H. Insley and V. D. Frechette, *Microscopy of Ceramics and Cements,* Academic Press, New York (1955).

W. D. Kingery, *Introduction to Ceramics,* John Wiley and Sons, New York (1960), Chapter 13.

W. C. McCrone, Jr., *Fusion Methods in Chemical Microscopy,* John Wiley and Sons, New York (1957).

MORE ADVANCED TEXTS:

Growth and Perfection of Crystals, ed. by R. H. Doremus, B. W. Roberts, and D. Turnbull, John Wiley and Sons, New York (1958).

PROBLEMS

6.1 Look up the differences between limestone and marble. Describe the microstructure you might expect in each.

6.2 The resolving power of an optical microscope is only about 0.1 micron. Explain how slip steps like those in Figure 6.4 can be resolved when the Burgers vector is only one or two Ångstroms long.

6.3 The c/a ratio of zinc is 1.85. Give the Miller indices of the most likely cleavage planes in Figure 6.5.

6.4 What property differences might you expect for oriented versus unoriented polystyrene?

6.5 Contrast the preferred orientation of a metal wire with the oriented structure produced by drawing a noncrystalline polymer.

6.6 Describe the differences between a multiphase material and a polycrystalline material. Why are multiphase and polycrystalline materials used much more than single phase materials and single crystals in engineering applications?

6.7 Assuming that the Si-O tetrahedra in asbestos have the same edge length as the Al-O octahedra, show why the double chains in the structure might spiral into hollow tubes rather than remaining flat.

6.8 On heating, pure iron undergoes an allotropic transformation from α (BCC) to γ (FCC) at 910°C. Does the iron expand, contract or retain its original volume during the $\alpha \rightarrow \gamma$ transformation? What are your assumptions?

6.9 Explain why lead and tin, unlike most metals, do not strain-harden when slowly deformed at room temperature.

6.10 Explain how the structure of a single crystal changes after it is (a) cold-worked and (b) cold-worked and annealed at a temperature well above that at which a cold-worked polycrystalline sample would have recrystallized.

6.11 The grain size of a copper sample can be refined (made smaller) only by heating it *after* it has been cold-worked. However, the grain size of a large-grained iron sample can be refined by heating it without the necessity for prior cold-work. Explain this difference in behavior.

6.12 Solid-state phase transformations often start at grain boundaries in the parent phase. Explain, in terms of atomic packing why this is the case.

6.13 Copper which is severely cold-worked is found to be 50% recrystallized at the following combinations of heating times and temperatures: (203°C, 6 sec); (162°C, 1 min); (127°C, 10 min); (97°C, 100 min); (72°C, 1000 min). The rate of recrystallization, like many other rate phenomena, is given by an equation of the form

$$\text{rate} = A \exp\left(\frac{-B}{T}\right)$$

where A and B are positive constants characteristic of the process and T is the absolute temperature. Taking the rate to be proportional to the reciprocal of the time for 50% recrystallization, estimate the time required for 50% recrystallization at −23°C.

6.14 The data for 50% recrystallization of one of the Saran copolymers are as follows: (10°C, 500 min); (20°C, 85 min); (40°C, 6 min); (80°C, 30 sec); (110°C, 0.2 sec). Test these data to see if they obey a rate equation of the type given in Problem 6.13, which describes the recrystallization behavior of metals.

6.15 Why is it that primary recrystallization is seldom observed in ceramic materials but that at elevated temperatures exaggerated grain growth is common?

6.16 A fine-grained, cold-worked rod $\frac{1}{4}$ inch in diameter is heated at a temperature where it recrystallizes and where grain growth proceeds at a rapid rate. When grain growth stops, it is found that the rod is *not* a single crystal. What grain structure is likely in the rod, and about what size are the grains when growth stops?

6.17 Grain size in a polycrystalline material may be specified (Timken–ASTM) by the *grain size number, n,* which appears in the equation

$$N = 2^{n-1}$$

where N is the number of grains per square inch of image at 100X (linear magnification). Some grains may appear quite small because the plane of sectioning barely intercepts them, while others may appear large because the plane passes through or near their centers. Statistical relationships have been established between the number of grains per unit area, (2^{n+3} grains/mm^2) on a random section, and the number of grains per unit volume ($2^{1.5n+4}$ grains/mm^3) in a sample of metal having grain size n. Consider the following materials, all of which have been produced to contain the same number of grains per mole (the densities follow each in parentheses): Mg (1.74), NaCl (2.165), Al (2.70), Al_2O_3 (3.97), TiB_2 (4.5), Zn (7.13), UO_2 (10.9), Pb (11.34), W (19.3).

(a) Compute the edge lengths, in inches, of cubes containing one mole of each of the above materials. Is there anything surprising about the results?

(b) What material in the above list has the largest grain size, and what one has the smallest grain size? If $n = 2$ for magnesium, what are the values of n for those two materials?

6.18 Estimate the grain size number and the number of grains per mm^2 for photomicrographs of nickel in Figure 6.12*c* and *d*.

CHAPTER SEVEN

Equilibrium Diagrams

Equilibrium diagrams are graphs that show which phases are present in a material at equilibrium with its surroundings. Properly interpreted, an equilibrium diagram shows the number of phases that are present, their compositions, and the relative amount of each as functions of temperature, pressure, and the over-all composition of the material. Although most engineering materials exist in a metastable, or nonequilibrium, state, any spontaneous change will be toward equilibrium, and much useful information about phase changes in such materials can be deduced from the appropriate equilibrium diagrams. Equilibrium diagrams are classified as unary, binary, and higher order (ternary, quarternary, and so forth) depending on the number of pure components involved. Of these, binary diagrams are used more extensively than the others and often are subdivided according to the invariant phase transformations which they contain.

7.1 INTRODUCTION

Phases in a material have been defined, in terms of microstructure, as regions that differ from one another either in composition or in structure, or both. For example, the state of aggregation (solid, liquid, or vapor), although one part of the structural description of a material, is not sufficient in all cases. A solid element, compound, or alloy may exist in any of several phases with different crystal structures. Also, a two-component material in its molten state may consist of two liquid phases with different compositions. In order to describe fully the structure of a material, we draw maps that show which phases are present and their relative amounts in a material, as functions of temperature, pressure and over-all composition. For a material in which each phase is at equilibrium with its surroundings, such a map is called an *equilibrium diagram;* otherwise it is called a *phase diagram.* A number of representative equilibrium diagrams are illustrated in this chapter and in Appendix VI.

7.2 THE PHASE RULE

It will be shown in Volume II (*Thermodynamics of Structure*) that thermodynamic considerations led Gibbs to derive a relationship between the number of phases (P) that can coexist at equilibrium in a given system, the minimum number of components (C) that can be used to form the system, and the degrees of freedom (F). The relationship may be stated in equation form as

$$P + F = C + 2 \tag{7.1}$$

which is known as the *Gibbs Phase Rule.* In this equation, the degrees of freedom are defined as the number of variables—temperature, pressure and composition—changes in which can be specified independently without changing the number of phases in equilibrium. There are $P(C - 1)$ composition variables for any system, since the concentration of every component but one must be specified to define the composition of each phase, and there are P phases.

7.3 UNARY DIAGRAMS

The variables that determine which phases exist in a material at equilibrium are temperature, pressure, and over-all composition. Clearly, in the case of a unary, or one-component, system, only temperature and pressure may be varied. The coordinates of unary equilibrium diagrams are therefore temperature and pressure. By convention, temperature usually is chosen as the abscissa, but since temperature is generally chosen as the ordinate in binary diagrams, we shall choose temperature as the ordinate here also.

The general form of the unary diagram (Figure 7.1) may be deduced from the phase rule. Consider the points representing equilibrium between two phases. Since the number of components is one, and the number of phases is two, one degree of freedom exists. This means that we can specify slight changes in either temperature or pressure, but not both, arbitrarily without changing the number of phases (two) in equilibrium. Once a temperature

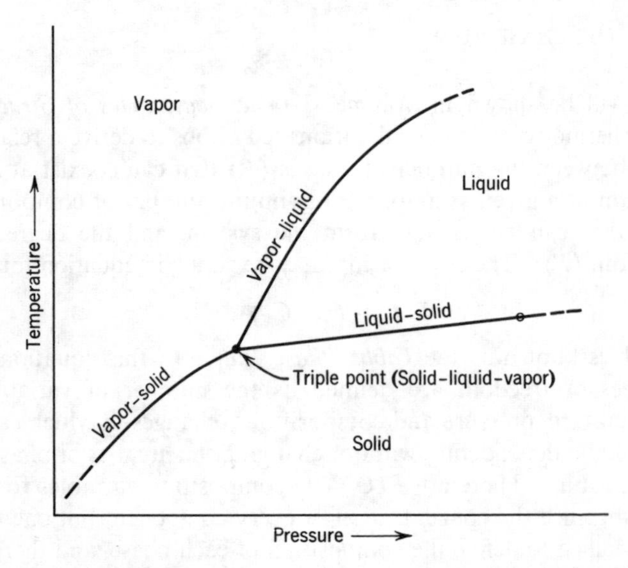

Figure 7.1 A hypothetical unary pressure-temperature diagram. The coordinates (P,T) define a point which lies either within the solid, liquid or vapor fields, or on a line bounding these fields. In the first case, the material will be single phase and will be in the state of aggregation specified by the field in which the point lies. In the second case, an equilibrium mixture of phases specified by the fields adjacent to the point can coexist.

is chosen, there is only one pressure at which the two phases will be in equilibrium. Thus two-phase equilibria are represented by lines or curves on a unary diagram. Suppose next that we wish to find the conditions under which solid, liquid, and vapor coexist at equilibrium. The number of components is still one; the number of phases is three; therefore no degrees of freedom exist. We cannot designate the values of any of the variables arbitrarily, and the three phases may coexist in equilibrium only at a specific temperature and a specific pressure. These particular values of temperature and pressure define a single point on the diagram which is called the *triple point.*

For many materials, the liquid-solid phase boundary is nearly horizontal, that is, the melting point is almost independent of pressure. The unary diagram for pure iron (Figure 7.2) has such

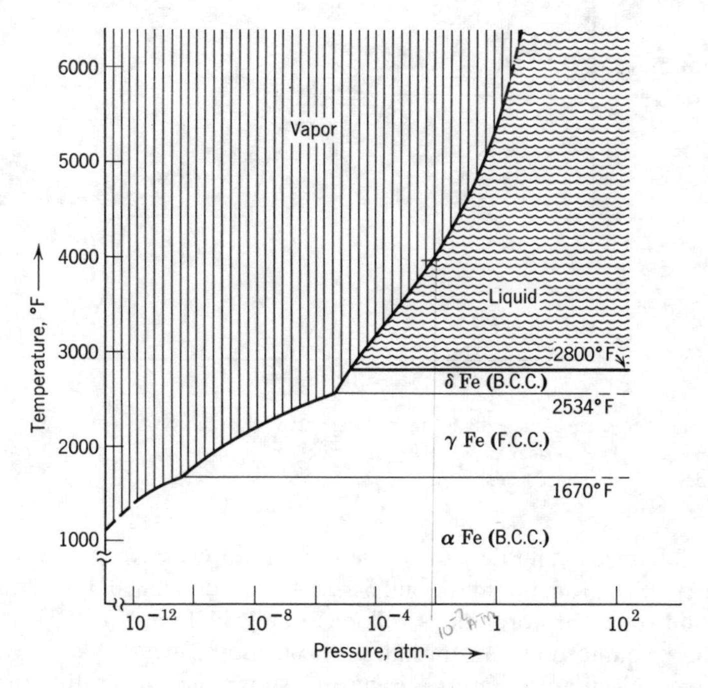

Figure 7.2 Approximate pressure-temperature diagram for pure iron.

a liquid-solid phase boundary. The field indicating the solid state in Figure 7.2 is subdivided into three regions: at 2534° F, two solid phases with different crystal structures can coexist; the same is true at 1670° F. Therefore three triple points are shown on this diagram, but only one of them represents equilibrium between a solid, a liquid, and a vapor phase. If a sample of liquid iron at one atmosphere pressure and 4000° F is cooled so slowly that equilibrium is

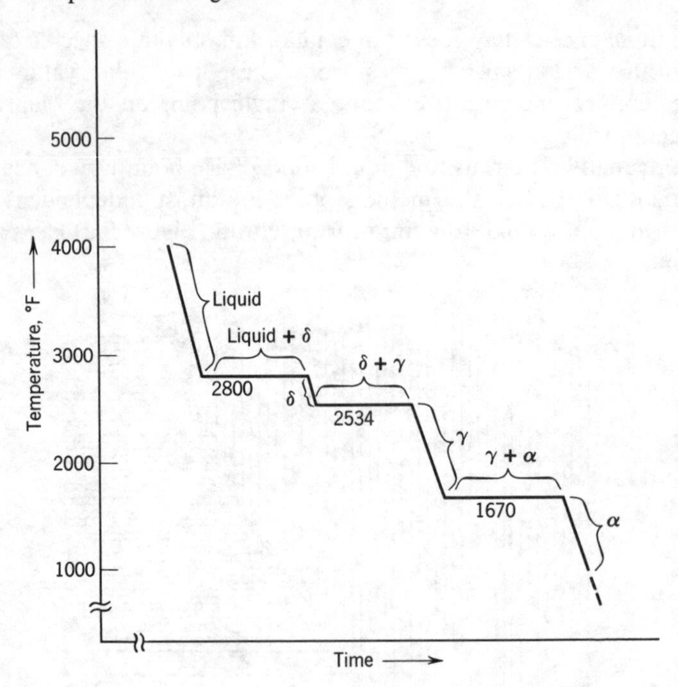

Figure 7.3 Schematic equilibrium cooling curve for pure iron at one atmosphere pressure. The thermal arrests occur at temperatures corresponding to the boundaries between phase fields in Figure 7.2.

maintained at all times, it may be seen from Figure 7.2 that a transformation from liquid to solid δ iron will occur at 2800° F and that solid state transformations will occur at 2534°F and 1670°F. All these equilibrium transformations occur *isothermally,* that is, with no change in temperature. Figure 7.3 shows the temperature-time curve for the equilibrium cooling of liquid iron at one atmosphere pressure. The curve is called a *cooling curve,* and abrupt changes in its slope signal the start or completion of a phase transformation.

7.4 TWO-COMPONENT SYSTEMS—SOLID SOLUBILITY

Two components are completely soluble in each other if the equilibrium state of any combination of the two is a single phase. Varying degrees of solubility occur: liquid water and alcohol are soluble in each other in all proportions at room temperature; they produce a homogeneous, single-phase liquid. Copper and nickel also are soluble in each other in all proportions, both in the liquid state and the solid state. The concept of a solid solution is not as familiar, in general, as that of a liquid solution, but the meaning is exactly the same; atoms or molecules of one component may be accommodated in the structure of the other component. A solid solution may be either *substitutional* or *interstitial:* the solute atom may substitute for a solvent atom or it may occupy an interstitial position between solvent atoms. Substitutional solid solutions usually are formed between two kinds of atoms of about the same size, and interstitial solid solutions usually are formed between atoms that differ greatly in size.

Only substitutional solid solutions can be formed in all proportions of both components. Certain conditions, known as the Hume-Rothery rules, must be satisfied if a solution of this type is to be formed. The two atoms must show

1. less than about 15 per cent difference in size
2. the same crystal structure
3. no appreciable difference in electronegativity
4. the same valence.

A solid equilibrium phase which forms from two components may be a solid solution. Or, if the Hume-Rothery rules are not satisfied, it may be a compound or an intermediate phase having a crystal structure different from either of the pure elements. No well-defined convention exists concerning the difference between a compound and an intermediate phase; in general, we shall consider a phase to be a compound only if it has a very limited range of solubility.

7.5 BINARY DIAGRAMS

A pair of pure elements or pure compounds may be mixed together in an infinite number of different proportions. For each over-all composition, the equilibrium state (the number of coexisting phases, their compositions and the relative amounts of each) is a function of temperature and pressure. Most operations used in the processing of materials are done at, or near, atmospheric pressure. Therefore the pressure often is not a significant variable and, in the diagrams which we shall consider, the pressure is specified at one atmosphere. Since one degree of freedom has been used in specifying the pressure, the phase rule now has the form:

$$P + F = C + 1 \tag{7.2}$$

In binary diagrams, temperature is plotted as the ordinate with composition as the abscissa. Since $C = 2$, by definition, one-phase equilibrium has two degrees of freedom—the temperature and composition of the phase—and is represented by an area, or a "phase field," in the equilibrium diagram. Two-phase equilibrium has one degree of freedom; if temperature is specified, the compositions of both phases in equilibrium are determined. Two-phase equilibrium is represented on a binary diagram by two lines, which are the temperature—composition curves for the two phases in equilibrium with each other. By similar reasoning, three-phase equilibrium is represented by a point; we say it is *invariant.*

A schematic *solid solution diagram* between two components, A and B, which are completely soluble in one another, is shown in Figure 7.4. The diagram is composed of a single-phase liquid region, a single-phase solid region, and a two-phase liquid plus solid region. The temperature-composition curves for each phase in two-phase equilibrium are the two curves that separate the single-phase from the two-phase regions. The *liquidus* is the temperature-composition curve for the liquid phase that is in equilibrium with solid; the *solidus* is the temperature-composition curve for the solid phase that is in equilibrium with liquid. The fact that liquidus and solidus do not coincide except at unique points (in this case, the melting points of A and B) verifies the phase rule which states that two-phase equilibrium must result in one degree of freedom. If solidus and liquidus coincided, changing the temperature of a two-phase mixture would change the number of phases in equilibrium from two to one.

Since the solidus and liquidus are temperature-composition curves for the two phases in equilibrium, the *ends* of a horizontal line drawn between the two curves will represent the compositions of the two phases at the temperature indicated by the horizontal line. Such horizontal lines in two-phase regions are called *tie lines.*

Using the preceding information, it is possible to determine from an equilibrium diagram the compositions of the phases that are

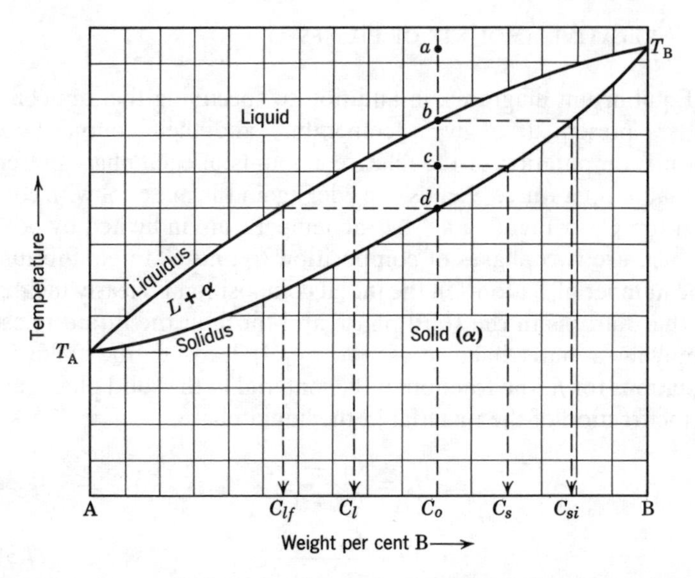

Figure 7.4 Binary solid solution diagram: an equilibrium diagram for elements A and B which are completely soluble in each other in all proportions in both the liquid and the solid state. Construction lines are for use in conjunction with the text. T_A and T_B are the melting points of pure A and pure B, respectively.

present at any temperature in a material of given over-all composition, provided the material is at equilibrium. Consider an alloy of composition C_o in the equilibrium diagram in Figure 7.4. If this alloy is at equilibrium at the temperature corresponding to point *a*, it is a single-phase liquid of composition C_o; if it is then cooled slowly to point *b*, the initial solid to form is of composition C_{si}. On further cooling (slowly enough to maintain equilibrium), the average composition of the solid follows the solidus, and the average composition of the liquid follows the liquidus until the temperature corresponding to point *c* is reached, at which the material consists of solid of composition C_s in equilibrium with liquid of composition C_l. Upon further cooling to point *d*, the final remaining liquid has the composition C_{lf}, and the over-all composition of the solid coincides with the over-all composition of the initial alloy. At any temperature below that corresponding to point *d*, the material is completely solid with composition C_o.

7.6 RELATIVE AMOUNTS OF PHASES

Equilibrium diagrams, in addition to specifying the number of phases present at a given temperature and their compositions, permit computation of the relative amounts of each phase present at that temperature. Let us consider again the material with composition C_o in Figure 7.4. At the temperature indicated by point *c*, there are two phases of composition C_l and C_s in equilibrium. The number of B atoms in the initial composition is clearly the sum of the B atoms in the solid phase and those in the liquid phase. Applying a mass balance (Problem 7.4) leads to the following equations for f_s, the fraction of the material in the solid phase, and f_l, the fraction of the material in the liquid phase.

$$f_s = \frac{C_o - C_l}{C_s - C_l} \tag{7.3a}$$

$$f_l = \frac{C_s - C_o}{C_s - C_l} \tag{7.3b}$$

These relationships, which are applicable in any two-phase region of a binary equilibrium diagram, are known as the *lever rule.* They are so called because a horizontal tie line within a two-phase region may be considered as a lever with fulcrum at C_o. The fraction of a phase having a composition indicated by one end of the lever is equal to the ratio of the length of the lever on the far side of the fulcrum to the total lever length.

7.7 THERMAL ANALYSIS

Equilibrium diagrams may be interpreted, as well as determined,[1] in terms of the cooling curves of different compositions. If pure copper, pure nickel, and a 50–50 Cu-Ni alloy are melted, and then are cooled at an extremely slow rate to maintain equilibrium, the cooling curves would look approximately like those illustrated in Figure 7.5*a*. Notice the similarity of these curves to the curve shown in Figure 7.3. The actual slopes are not of great significance,

[1] For other methods of determining phase diagrams, see *Phase Diagrams in Metallurgy* (McGraw-Hill, 1956) by F. N. Rhines.

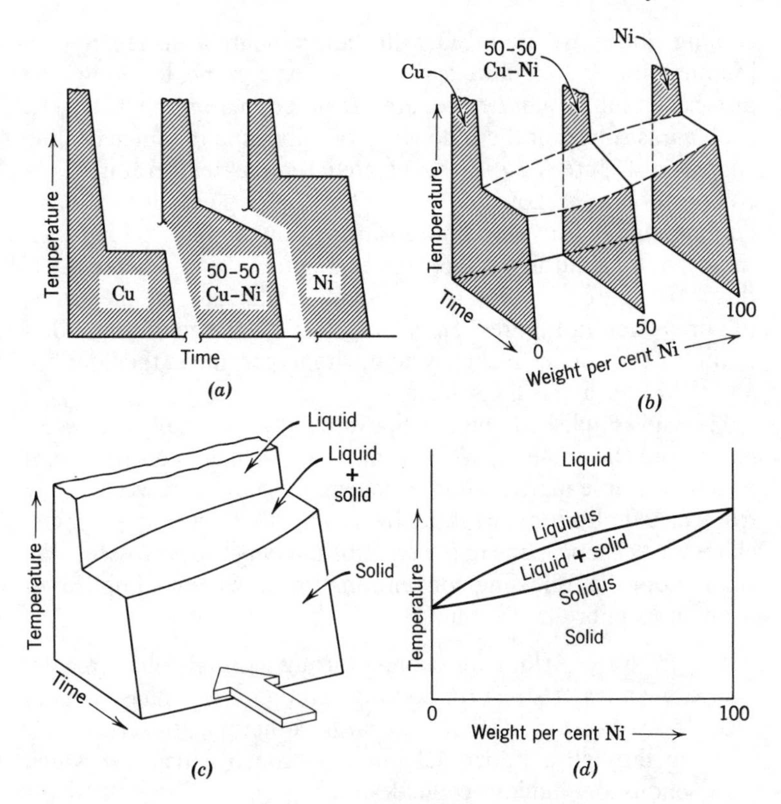

Figure 7.5 Relationship between cooling curves and equilibrium diagram for the system copper-nickel: (*a*) Individual cooling curves for Cu, Ni and a 50-50 Cu-Ni alloy. (*b*) The cooling curves of (*a*) on time-temperature-composition coordinates. (*c*) Surface generated by cooling curves for all possible alloys of Cu with Ni. (*d*) Cu-Ni equilibrium diagram, which is the surface in (*c*) as viewed in the direction indicated by the arrow.

for they can be altered by changing the rate of heat removal, but the temperatures at which abrupt changes occur are significant. They indicate the start or completion of a phase transformation, or phase change. The curves also illustrate the fact that each pure component solidifies at a constant temperature, but a multicomponent material often solidifies over a range of temperatures.

Different compositions, of course, have different cooling curves. Therefore, in attempting to construct an equilibrium diagram from cooling curves, we are faced with the problem of representing an infinite number of cooling curves on a single graph, for an infinite number of alloys can be prepared from copper and nickel. The problem is solved in the following way: imagine the three cooling curves in Figure 7.5*a* to be erected on the temperature-time-composition axes shown in Figure 7.5*b*. A large number of other cooling curves for other compositions may be inserted between these until a solid figure like that in Figure 7.5*c* is generated. If this solid figure is viewed in the direction indicated by the arrow, the projection in Figure 7.5*d* is seen. This is the equilibrium diagram for the copper-nickel system; the upper line is the liquidus, and the lower line is the solidus.

Having established the relationship between equilibrium diagrams and the cooling curves from which they are constructed, it is now a simple matter to draw schematic cooling curves directly from an equilibrium diagram by noting that their slopes must change every time a line in the equilibrium diagram is crossed. We shall adopt the following conventions for drawing cooling curves from an equilibrium diagram:

1. draw a steep slope for cooling through a single-phase region.
2. draw a lesser slope for cooling through a two-phase region.
3. draw a horizontal slope (indicating a thermal arrest) for cooling through a horizontal line or through a point at which solidus and liquidus coincide.

Figure 7.6 illustrates, schematically, how a cooling curve can be extracted from an equilibrium diagram.

7.8 LIMITED SOLID SOLUBILITY

Pairs of components which fulfill all the Hume-Rothery rules (Section 7.4) are not particularly common. If any of these rules is not satisfied, two or more solid phases will exist in the equilibrium diagram. As an example, suppose that component A has a FCC crystal structure and component B has a BCC crystal structure. As B atoms are substituted for A atoms in the FCC structure, the lattice becomes somewhat distorted compared to that of pure A, but it is still recognizable as FCC. At some particular composi-

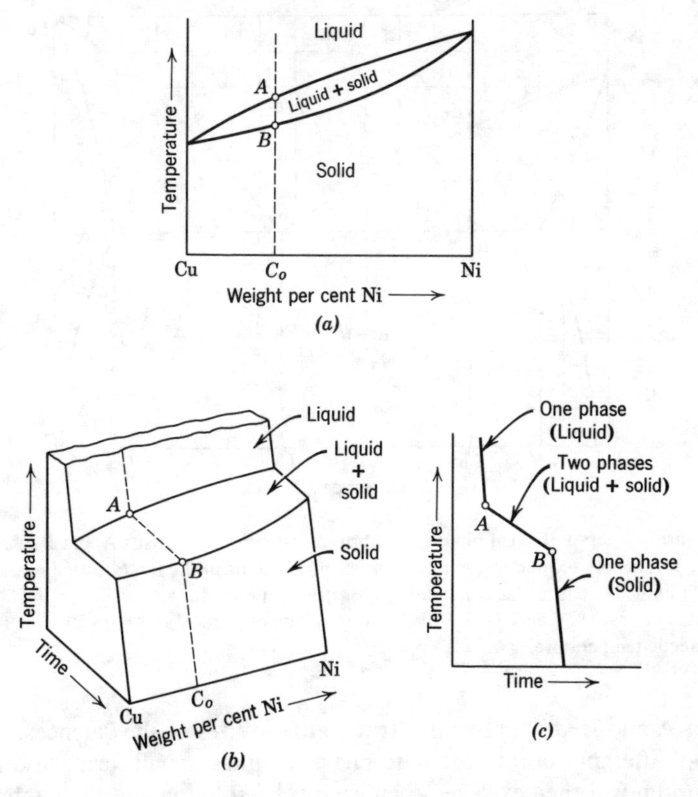

Figure 7.6 Schematic cooling curves for Cu-Ni alloy of over-all composition C_o. Points *A* and *B* correspond in all three illustrations. (*a*) Location of C_o on the equilibrium diagram. (*b*) Trace of cooling curve on three-dimensional surface. (*c*) Cooling curve of alloy of composition C_o.

tion, however, the energy of this substitutional solid solution becomes so great that the addition of more B atoms causes the single phase to separate into two phases: a FCC phase, designated α, consisting primarily of A atoms with some B atoms in solid solution; and a BCC phase, designated β, consisting primarily of B atoms with some A atoms in solution.

Even if both kinds of atoms have the same crystal structure, if one is electronegative and the other electropositive, an intermediate phase (often a compound showing a very limited range of solubility

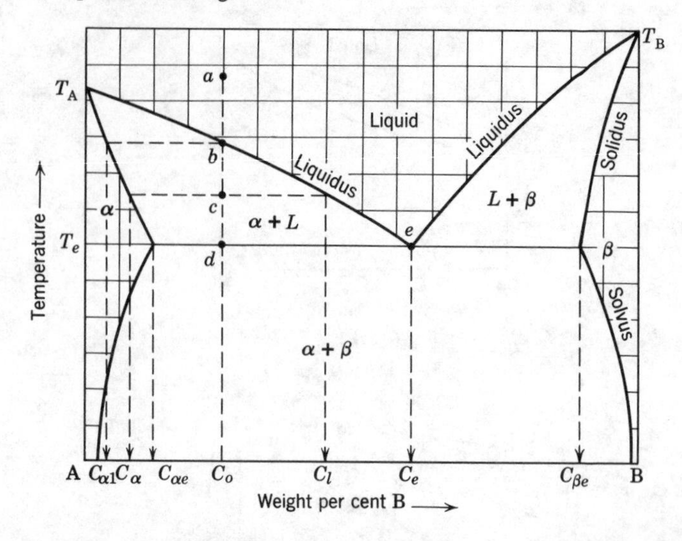

Figure 7.7 Hypothetical binary equilibrium diagram for elements A and B which are completely soluble in each other in all proportions in the liquid state but only to a limited extent in the solid state. Construction lines are for use in conjunction with the text. T_A and T_B are the melting points of pure A and pure B; T_e is the eutectic temperature.

for A and B atoms) forms. If the atomic sizes or the valences are too different, an intermediate phase, or phases, will form. Equilibrium will then exist between each solid solution and the nearest intermediate phase.

7.9 THE BINARY EUTECTIC DIAGRAM

One of the types of equilibrium diagrams which may result when there is only limited solubility in the solid state is a *binary eutectic diagram,* illustrated schematically in Figure 7.7. Consider alloy C_o, which exists as a single-phase liquid at point *a*; when it is cooled to point *b*, the composition of the first solid to form is given by the other boundary of the two-phase region, $C_{\alpha 1}$. On further cooling to point *c*, a solid phase of composition C_α and a liquid of composition C_l are at equilibrium. As in the case of the solid solution diagram, the relative amounts of the two phases in equilibrium may be calculated by the lever rule. At point c, the fraction which is α phase is $(C_l - C_o)/(C_l - C_\alpha)$, and the fraction which is liquid phase is $(C_o - C_\alpha)/(C_l - C_\alpha)$. A similar situation existed initially in the solid solution diagram considered in Section 7.5. In fact, both of the upper two-phase regions, $(\alpha + L)$ and $(L + \beta)$ in Figure 7.7 can be considered as parts of a pair of solid solution diagrams, one for the solution of A in B and the other for the solution of B in A.

If the material is cooled still further below point c, more solid forms, and the composition of the liquid follows the liquidus down to point e, which is called the *eutectic point.* With further extraction of heat, the *eutectic liquid* of composition C_e solidifies *isothermally* at the *eutectic temperature* T_e. This phase transformation is called an *eutectic reaction.* It is an invariant of the system; since three phases are in equilibrium during solidification of the eutectic liquid, there are no degrees of freedom. The temperature, the composition of the liquid phase, and the compositions of both solid phases are fixed.

Once the eutectic liquid has solidified completely, the temperature again starts to decrease. It may be seen from Figure 7.7 that the solubility of B in A (and of A in B) decreases with decreasing temperature below T_e; the excess B precipitates on cooling, and the composition of the phase follows the *solvus* or maximum solubility line. The excess B atoms which are no longer soluble in the α at the lower temperature are not rejected as pure B because, as indicated in the diagram, A is soluble to a certain extent in B; the precipitate is actually β.

The solid state microstructure of a material having composition C_e in Figure 7.7 will be an intimate mixture of two phases. The α and β phases in such an eutectic material may be in the form of thin (of the order of a micron) plates or rods or tiny particles. The lamellar microstructure of a lead-tin eutectic is shown in Figure 7.8*a*. A material with composition between $C_{\alpha e}$ and C_e is called *hypoeutectic* and, in general, will have a microstructure containing *primary* α (α formed above T_e) in a matrix of eutectic. A material with composition between C_e and $C_{\beta e}$ is called *hypereutectic* and, in general, will have a microstructure containing primary β in a matrix of eutectic. A microstructure showing primary tin solid solution in a matrix of lead-tin eutectic is shown in Figure 7.8*b*.

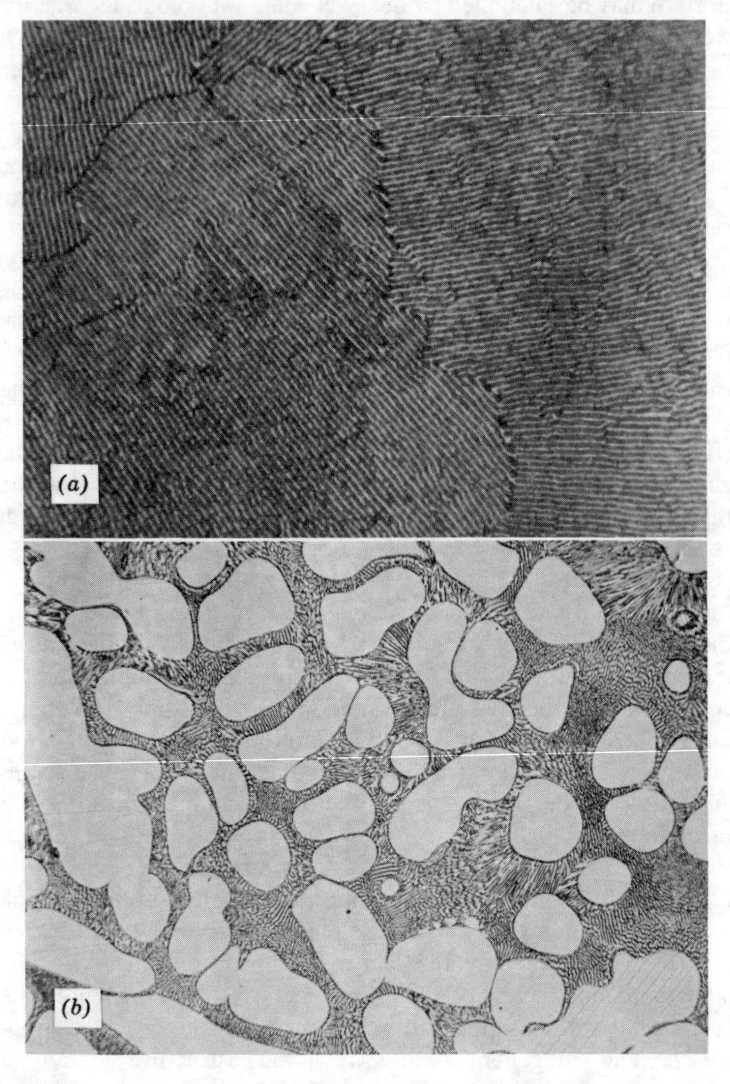

Figure 7.8 Microstructures of two lead-tin alloys. (*a*) An eutectic composition alloy, showing the intimate mixture of the two phases, tin solid solution (light) and lead solid solution (dark). 500X. (*b*) A hypereutectic lead-tin alloy, showing primary tin solid solution in a matrix of lead-tin eutectic. 100X.

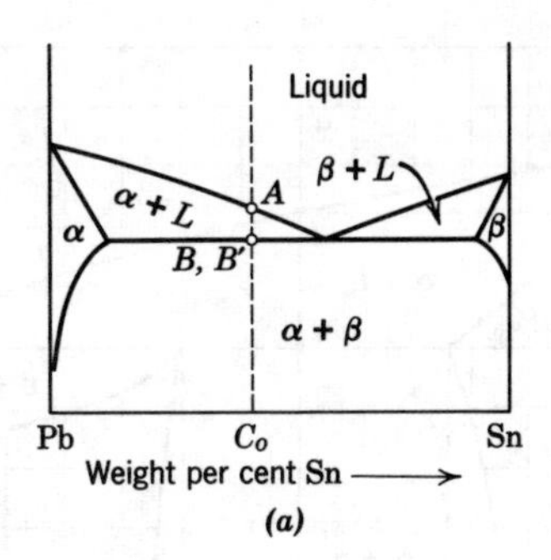

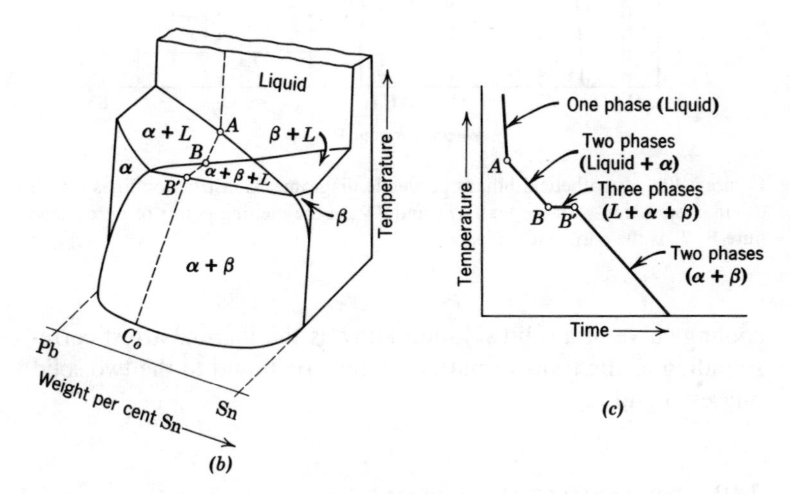

Figure 7.9 Schematic cooling curves for Pb-Sn alloy of over-all composition C_o. Points A, B and B' correspond in the three illustrations. (*a*) Location of C_o on the equilibrium diagram. (*b*) Trace of cooling curve on three-dimensional surface. (*c*) Cooling curve of alloy of composition C_o.

In the few cases where the α phase which forms from the eutectic reaction crystallizes directly on primary α so that the matrix is all β rather than an eutectic mixture, the eutectic is called a *divorced eutectic.*

The relationship between a simple eutectic equilibrium diagram and a cooling curve for a hypoeutectic lead-tin alloy is demonstrated in Figure 7.9. The additional feature, compared to the

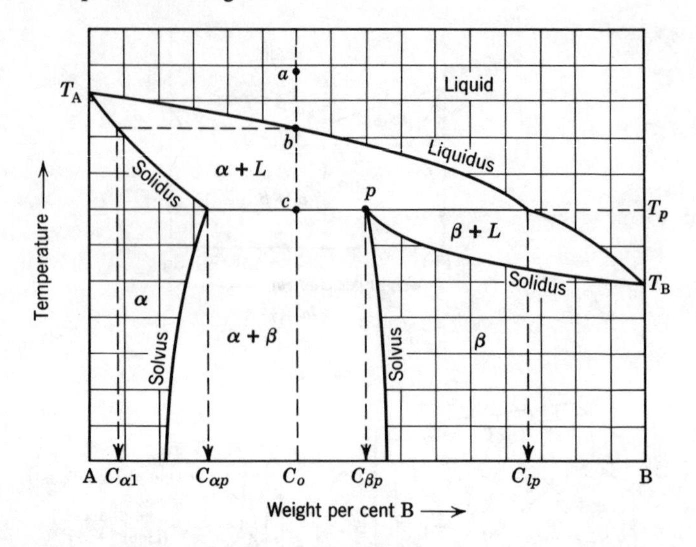

Figure 7.10 Hypothetical binary peritectic diagram. Construction lines are for use in conjunction with the text. T_A and T_B are the melting points of pure A and pure B; T_p is the peritectic temperature.

cooling curve of a solid solution alloy, is the thermal arrest corresponding to the transformation of eutectic liquid to the two solid phases, α and β.

7.10 THE PERITECTIC DIAGRAM

A third kind of simple binary equilibrium diagram is the *peritectic diagram,* shown schematically in Figure 7.10. It contains three two-phase regions, just as the eutectic diagram did, and the lever rule is again applicable in each of these regions. In a typical peritectic reaction, a material of over-all composition C_o existing as a liquid at point *a* transforms to $\alpha + L$ during cooling as in a solid solution or in a eutectic diagram. But, at the *peritectic temperature* T_p, α of composition $C_{\alpha p}$ and liquid of composition C_{lp} react to produce β of composition $C_{\beta p}$. Application of the lever rule to the $\alpha + \beta$ field just below T_p in Figure 7.10 shows that all the α is not consumed by the peritectic reaction unless the over-all composition C_o of the material is equal to or greater than $C_{\beta p}$. The fraction which is α just below T_p is $(C_{\beta p} - C_o)/(C_{\beta p} - C_{\alpha p})$. At the *peritectic composition* $C_{\beta p}$, *all* the α and *all* the liquid co-existing just above T_p are consumed to produce a single phase, β. Note that in this schematic diagram (which is typical of many peritectics), after the peritectic reaction the composition of the resulting β follows the solvus line and an α precipitate is formed within the β.

The peritectic reaction, like the eutectic reaction, is an invariant of the system. Examples of simple peritectic equilibrium diagrams are rare, but the peritectic reaction is quite common as a part of more complicated equilibrium diagrams, especially when the melting points of the two components are very different.

7.11 INVARIANT REACTIONS

Equilibrium diagrams of actual binary systems are usually not simple solid solution, eutectic, or peritectic diagrams. Instead they are composite diagrams containing a number of two-phase regions and a number of invariant reactions. The more common invariant reactions have been given specific names and are of two general kinds.

1. On cooling, one phase separates into two phases.
2. On cooling, two phases react to produce a third, different phase.

It is conventional to describe these reactions in equation form, where L represents a liquid phase and a Greek letter represents a solid phase (including compounds). Those invariant reactions that belong to class (1) are

a. Monotectic: $L_1 \xrightarrow{\text{cooling}} \alpha + L_2$
b. Eutectic: $L \xrightarrow{\text{cooling}} \alpha + \beta$
c. Eutectoid: $\gamma \xrightarrow{\text{cooling}} \alpha + \beta$

Those belonging to class (2) are

a. Syntectic: $L_1 + L_2 \xrightarrow{\text{cooling}} \beta$
b. Peritectic: $\alpha + L \xrightarrow{\text{cooling}} \beta$
c. Peritectoid: $\alpha + \gamma \xrightarrow{\text{cooling}} \beta$

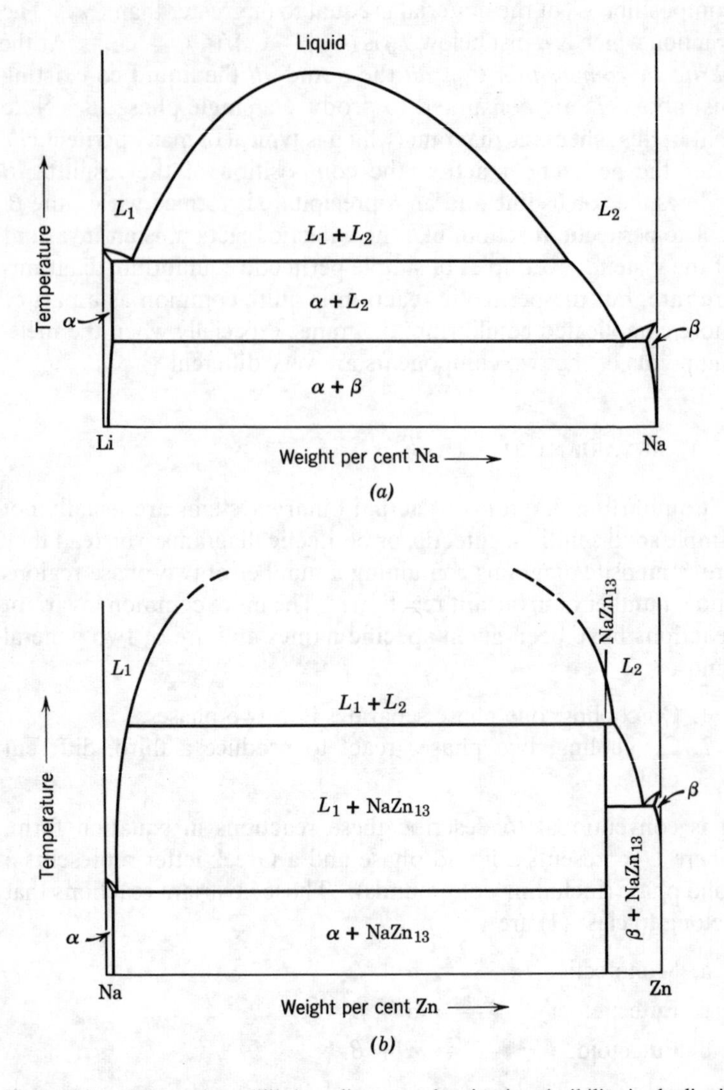

Figure 7.11 Schematic equilibrium diagrams showing immiscibility in the liquid state over certain temperature ranges. (*a*) Monotectic. (*b*) Syntectic. The α and β fields of both diagrams actually are vanishingly narrow but are indicated with finite width for clarity.

Figure 7.11 illustrates the monotectic reaction in the Li-Na system and the syntectic reaction in the Na-Zn system.

7.12 EQUILIBRIUM DIAGRAMS WITH MORE THAN ONE INVARIANT REACTION

The more complex equilibrium diagrams often show intermediate phases, or compounds, as well as more than one invariant reaction. Figure 7.12 shows the equilibrium diagrams for the systems Mg-Pb and Sb-Te, each of which consists of two adjacent eutectics with an intermediate phase. In the Mg-Pb system the intermediate phase is a compound, Mg_2Pb; neither excess magnesium nor excess lead is soluble in the compound. On the other hand, in the Sb-Te system the phase θ with nominal composition Sb_2Te_3 has a fairly wide range of solubility except at its melting point.

The Au-Pb equilibrium diagram, which is shown in Figure 7.13*a*, contains two intermediate compounds Au_2Pb and $AuPb_2$. Each forms on cooling by a peritectic reaction. On the other hand, the peritectic reaction in the Hg-Pb system, Figure 7.13*b*, results in an intermediate phase of nominal composition $HgPb_2$, which shows a range of solubility except at the peritectic point.

In spite of the apparent complexity of many equilibrium diagrams, they may be used exactly as the simpler diagrams. For a given over-all alloy composition, a tie line across a two-phase region gives the compositions of the two phases in equilibrium; the lever rule may be used to compute the relative amounts of each phase and isolation of an invariant reaction permits the determination of the three phases in equilibrium at that temperature.

DEFINITIONS

Alloy: A multicomponent solid or liquid in which the primary component is a metal.

Compound: A multicomponent phase which exists over a very narrow range of compositions.

Cooling Curve: A plot of the temperature of a material versus time as it undergoes one or more phase transformations.

Equilibrium: A state in which no changes take place with time.

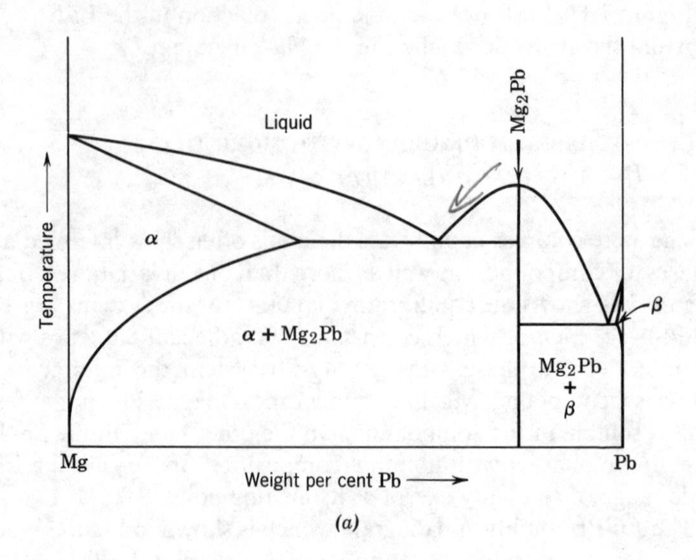

(a)

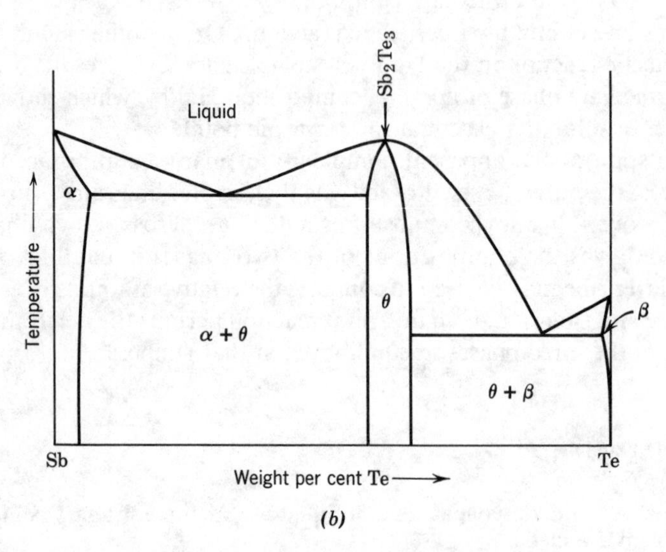

(b)

Figure 7.12 Schematic equilibrium diagrams showing two eutectic transformations separated (*a*) by an intermediate compound and (*b*) by an intermediate phase with a range of solubility.

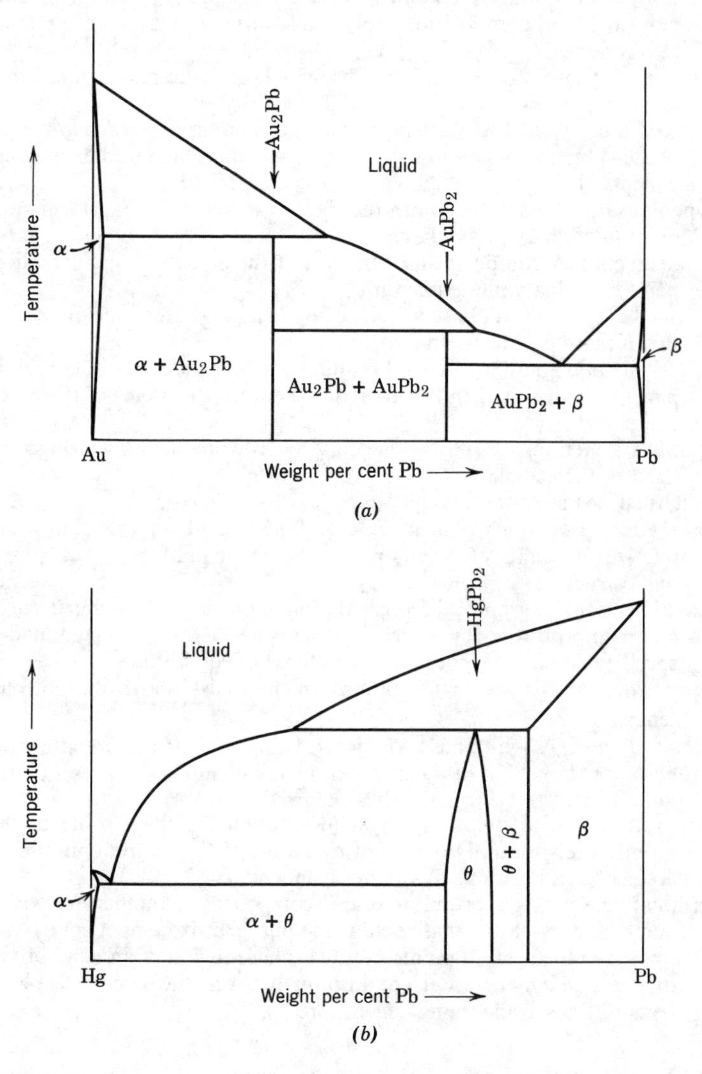

Figure 7.13 Schematic equilibrium diagrams showing peritectic transformations resulting in (*a*) the formation of a compound and (*b*) an intermediate phase with a range of solubility. In actuality, the α and β fields in the Au-Pb diagram and the α field in the Hg-Pb diagram have more limited solubility than is indicated here.

Equilibrium Diagram: A graphical representation of the pressures, temperatures and compositions for which various phases are stable at equilibrium.

Eutectic: A phase transformation in which all the liquid phase transforms on cooling to two solid phases simultaneously.

Gibbs Phase Rule: The statement that, at equilibrium, the number of phases plus the degrees of freedom must equal the number of components plus two.

Hypereutectic: A composition to the right of the eutectic transformation point in a binary phase diagram.

Hypoeutectic: A composition to the left of the eutectic transformation point in a binary phase diagram.

Intermediate Phase: A phase which occurs in a range of compositions between those of the terminal phases.

Interstitial Solid Solution: A solid solution in which the atoms of one component are dissolved in the interstices between the atoms of the other component.

Invariant Reactions: Equilibrium phase transformations involving zero degrees of freedom.

Isothermal: At a constant temperature.

Lever Rule: Equations such as 7.3*a* and 7.3*b*, used for calculating the relative amount of each phase in an equilibrium two-phase microstructure.

Phase Diagram: A graphical representation of the pressures, temperatures and compositions for which various phases are predicted under specified conditions, not necessarily those of equilibrium.

Proeutectic: A phase transformation which occurs above the eutectic temperature.

Primary Phase: A solid phase which forms in a microstructure at a temperature above that of an invariant reaction, and is still present after the invariant transformation has occurred.

Substitutional Solid Solution: A solid solution in which some of the atomic sites normally occupied by atoms of the matrix component are occupied by atoms of another component.

Terminal Phase: A solid solution of one component in another, for which one boundary of the phase field is the pure matrix component.

Tie Line: The locus of all points in a two-phase region of an equilibrium diagram which represent equilibrium between the same two phase compositions at the same temperature.

BIBLIOGRAPHY

Supplementary References:

M. Hansen, *Constitution of Binary Alloys,* McGraw-Hill Book Co., N. Y. (1958); a compendium of alloy phase diagrams.

E. M. Levin, H. F. McHurdie and F. P. Hall, *Phase Diagrams for Ceramists,* American Ceramic Society, Columbus (1956); a compendium of ceramic phase diagrams.

MORE ADVANCED TEXTS:

A. N. Campbell and N. O. Smith, *The Phase Rule and its Applications,* Dover Publications, N. Y. (1951).

F. N. Rhines, *Phase Diagrams in Metallurgy,* McGraw-Hill Book Co., N. Y. (1956).

PROBLEMS

7.1 A sample of iron is contained in a crucible within a tank made of an inert material. The tank initially contains argon at one atmosphere pressure and is fitted with a relief valve so that the total pressure always remains one atmosphere. If the entire tank is heated to 4000°F and held at this temperature long enough for the iron to heat up to temperature, what will be the pressure of argon in the tank? Suppose that a tank which is similar but lacks a relief valve is kept continually pumped down to a pressure of 10^{-4} atmosphere while maintaining it at a temperature of 4000°F. In what way will the result differ from that in the first case? Why?

7.2 Justify the convention, used in this chapter, of showing the cooling rate as lower through a two-phase region than through a single phase region, if heat is extracted at a slow, uniform rate.

7.3 Plot the Cu-Ni diagram from the following data:

Weight % Ni	*Liquidus Temperature, °C*	*Solidus Temperature, °C*
0	1083	1083
20	1195	1135
40	1275	1205
60	1345	1290
80	1410	1375
100	1453	1453

In what region or regions of the diagram can the lever rule be used?

7.4 An alloy of A and B of over-all composition C_o is at a temperature at which it is an equilibrium mixture of two phases α and β, of compositions C_α and C_β. Make a mass balance to derive the lever rule.

7.5 "Derive" the lever rule by treating the tie line as a mechanical lever, in balance, having a fulcrum at C_o with the weight of the α phase hanging on the α end and the weight of the β phase hanging on the β end of the lever.

7.6 Can the lever rule be applied at the temperature of an invariant reaction? Why or why not?

7.7 Determine the approximate composition of the alloy shown in Figure 7.8*b*. Assume that the area fraction of a phase equals the volume fraction. Take the density of the lead-rich phase to be 10.3 grams/cm^3 and the density of the tin-rich phase to be 7.3 grams/cm^3. The relative amounts of the phases are determined essentially by the eutectic reaction: L (61.9 wt.% Sn) $\xrightleftharpoons{183°\text{C}}$ α (19.2 wt.% Sn) + β (97.5 wt.% Sn).

7.8 Refer to Figure 7.7. At a temperature just below T_e, the weight fraction of an alloy of over-all composition C_o which exists as the α component of the eutectic phase is given by

$$\left[\frac{(C_{\beta e} - C_o)}{(C_{\beta e} - C_{\alpha e})}\right] - \left[\frac{(C_e - C_o)}{(C_e - C_{\alpha e})}\right]$$

Show that the weight fraction of C_o which exists as the α component of the eutectic is also given by

$$\left[\frac{(C_o - C_{\alpha e})}{(C_e - C_{\alpha e})}\right] \times \left[\frac{(C_{\beta e} - C_e)}{(C_{\beta e} - C_{\alpha e})}\right]$$

and prove that the two expressions are mathematically equivalent.

7.9 Refer to Figure 7.7. Write general expressions for the fraction of α that is proeutectic, the fraction that is eutectic and the fraction that is a precipitate in an alloy of over-all composition C_o at the lowest temperature shown. Sketch the expected microstructure for the alloy, showing the three forms of α and their most probable locations.

7.10 The invariant reaction α_1 (78 wt.% Zn) $\xrightleftharpoons{275°\text{C}}$ α (31.6 wt.% Zn) + β (99.4 wt.% Zn) occurs in the Al-Zn system. Name the invariant reaction. For a 58.4 wt.% Zn alloy, calculate the weight fractions of α and α_1 at 275.1°C, and calculate the weight fractions of α and β at 274.9°C.

7.11 The eutectic reaction in the Cr-Ni system (Appendix VI) is L (49 wt.% Ni) $\xrightleftharpoons{1345°\text{C}}$ α (35 wt.% Ni) + β (53 wt.% Ni). Consider two alloys, C_o and C_1, containing, respectively, less nickel and more nickel than the eutectic composition. The weight fraction of the proeutectic phase is the same for both, but immediately after completion of the eutectic reaction, the weight fraction of total α for alloy C_o is $2\frac{1}{2}$ times the weight fraction of α for alloy C_1. Calculate the compositions C_o and C_1 in weight percent Ni.

7.12 Two eutectic transformations in the Mg-Pb system (Figure 7.12*a*) are

$$L\ (66.8\text{ wt.\% Pb}) \rightleftharpoons \alpha\ (41.7\text{ wt.\% Pb}) + Mg_2Pb$$
$$L\ (97.8\text{ wt.\% Pb}) \rightleftharpoons Mg_2Pb + \beta\ (99.3\text{ wt.\% Pb})$$

Immediately upon completion of the eutectic reaction, in a 20 wt.% Mg alloy,

(a) What wt.% of the alloy is proeutectic Mg_2Pb?

(b) What wt.% of the alloy is Mg_2Pb which is contained in the eutectic solid?

(c) What is the ratio of the weight of Mg_2Pb contained in the eutectic to the weight of eutectic?

7.13 The system Cr-Pd contains an intermediate phase, Cr_xPd_y, having a composition 57.7 wt.% Pd and melting at 1398°C. Between Cr and Cr_xPd_y there is a eutectic reaction

$$L\ (40.6\ \text{wt.\%}\ \text{Pd}) \xrightleftharpoons{1320^\circ\text{C}} \alpha\ (10\ \text{wt.\%}\ \text{Pd}) + Cr_xPd_y.$$

Between Cr_xPd_y and Pd, there is a continuous series of solid solutions (β). The ($\beta + L$) field shows a minimum similar to the Cu-Au system (Appendix VI) at 62 wt.% Pd and 1295°C.

(a) Sketch the Cr-Pd equilibrium diagram and label all fields.

(b) Calculate the ratio x/y and assign a formula to the compound Cr_xPd_y in which x and y are small whole numbers.

(c) What is the likely crystal structure of Cr_xPd_y? Why?

7.14 The binary Re-Pt diagram is a simple peritectic diagram containing the reaction

$$L\ (57\ \text{wt.\%}\ \text{Pt}) + \alpha\ (46\ \text{wt.\%}\ \text{Pt}) \xrightleftharpoons{2450^\circ\text{C}} \beta\ (54\ \text{wt.\%}\ \text{Pt}).$$

Re is soluble in Pt to the extent of 40 wt.% at 1800°C, and Pt is soluble in Re to the extent of 43 wt.% at the same temperature. Draw a schematic cooling curve for a 54 wt.% Pt alloy slowly cooled from the liquid state and label each segment of the curve with the phases present. How do you account for the experimental observation that an alloy of peritectic composition (54 wt.% Pt) contains some α after it has cooled to room temperature?

7.15 (a) For the system Au-Pb (Figure 7.13*a*), calculate the ratio of the weight fraction lead-poor phase to lead-rich phase for an alloy of over-all composition 50 wt.% Pb.

(b) For the system MgO-SiO_2 (Appendix VI), calculate the ratio of the weight fraction SiO_2-poor phase to SiO_2-rich phase for an alloy of over-all composition 50 wt.% SiO_2.

7.16 In the Cu-Zn system (Appendix VI) identify all invariant reactions involving the liquid and write equations for them.

7.17 For the following metals (atomic radii are given in Å in parentheses after each metal) look up the crystal structures in Appendix II: Co (1.250), Ni (1.243), Pb (1.747), Pt (1.384), Ta (1.427), Ti (1.455), W (1.367), Zn (1.329). Co is HCP up to 390°C and FCC above that temperature; Ti is HCP up to about 900°C and BCC above that temperature. In the absence of any other information, which of the following pairs of metals would *not* be likely to form a continuous series of solid solutions: (W-Ta), (Pt-Pb), (Co-Ni), (Zn-Co), (Ti-Ta)? Check your predictions by referring to a tabulation of phase diagrams.

7.18 A congruent phase transformation is one in which one phase transforms directly into another without a change in composition. For instance, pure metals melt congruently. Classify the following as congruently or incongruently melting compositions:

(a) $MgO \cdot Al_2O_3$ (Appendix VI)
(b) 80% Au alloy of Cu-Au system (Appendix VI)
(c) θ phase of Sb-Te system (Figure 7.12*b*)
(d) $HgPb_2$ (Figure 7.13*b*)
(e) $2MgO \cdot SiO_2$ in MgO-SiO_2 diagram (Appendix VI)
(f) $NaZn_{13}$ (Figure 7.11*b*)

7.19 Explain why complete solubility can occur between the two components of a substitutional solid solution but not for an interstitial solid solution.

7.20 The room temperature microstructure of a MgO-Al_2O_3 ceramic (Appendix VI) shows uniform equiaxed grains which contain within them about 5 wt.% of a second phase in the form of a fine needle-like precipitate. If this is an equilibrium structure, estimate the over-all composition of the ceramic.

CHAPTER EIGHT

Nonequilibrium Phase Transformations

Many phase changes which occur in engineering materials do not result in equilibrium structures. Such phase changes are called nonequilibrium phase transformations. They are important, sometimes because the structures they produce are undesirable and sometimes because the structures have excellent engineering properties. Coring and surrounding are two nonequilibrium structures which result from cooling too rapidly for solid state diffusion to establish phase equilibrium. Cooling too rapidly also may suppress completely the formation of equilibrium phases; in some alloys, the result is simply a supersaturated solid solution at low temperatures, but in other alloys, such as the steels, completely new nonequilibrium phases form. The control of nonequilibrium phase transformations is one of the principal means of manipulating the engineering properties of solids.

8.1 INTRODUCTION

Although equilibrium diagrams specify what phases are present in a microstructure at equilibrium, as well as the composition and relative amount of each phase, they do not show the metastable nonequilibrium phases, of different compositions and in different amounts than the equilibrium phases, which are often present in a material. Equilibrium diagrams are, nonetheless, convenient reference maps from which to consider departures from equilibrium. Nonequilibrium structures are generally produced by cooling the material so rapidly that there is not enough time or thermal energy for the atoms to be rearranged in an equilibrium configuration. The noncrystalline solids discussed in Chapter 5 are one type of nonequilibrium structure. In alloys and in inorganic solids with mobile atoms, crystalline nonequilibrium phases are more common; these microstructures and the reasons for their existence are discussed in the following sections from an experimental point of view. A thermodynamic and kinetic rationale for this point of view is presented in Volume II (*Thermodynamics of Structure*) of this series.

8.2 CORING DURING SOLIDIFICATION

If a molten binary alloy solidifies through a liquid + solid region under equilibrium conditions, the compositions of the liquid and solid phases must readjust continuously as the temperature is lowered. Such readjustments are effected by the diffusion of both atomic species in both phases. But since the diffusion rate in the solid state tends to be slow, an extremely long time may be required to even out the composition gradients. In practice, cooling rates are almost always so rapid that the composition gradients remain; such a microstructure is said to be *cored* because the first regions to solidify (the "cores") have compositions different from those of the last material to solidify. Since a chemical etch often attacks regions of different compositions at different rates, cored regions can be delineated in a microstructure. A cored copper-nickel alloy is shown in Figure 8.1, and a cored aluminum-copper alloy is shown in Figure 6.11*b*. Figure 8.2 shows the process by which a cored structure forms. Consider a molten alloy of over-all composition C_o at temperature T_o; as it is cooled, the first solid to form has composition α_1. We assume that the solid forming at the solid-liquid interface at temperatures T_2, T_3, and T_4 has compositions α_2, α_3, and α_4, that is, that its composition is given by the equilibrium solidus. If the cooling rate is so rapid that each increment of solid formed maintains its initial composition, we may picture the *average* composition of all solid formed proceeding along a "nonequilibrium solidus" from α_1, to α_2' to α_3' and so on. The last liquid disappears only when the average composition of the solid is the same as the over-all composition of the alloy, that is, when the nonequilibrium solidus crosses the vertical line at C_o.

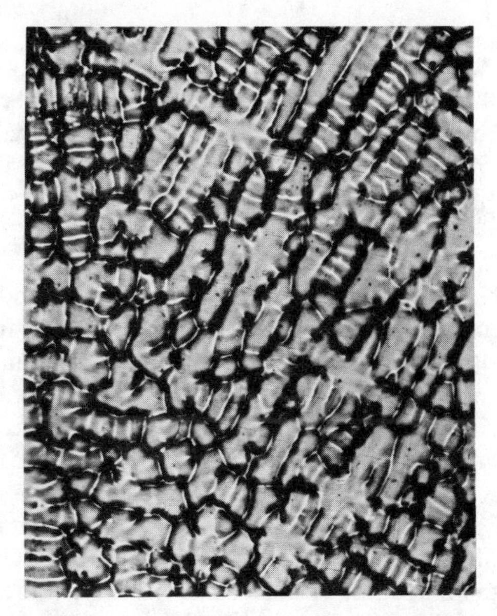

Figure 8.1 The microstructure of a cored, cast 70-30 Cu-Ni alloy. 75X.

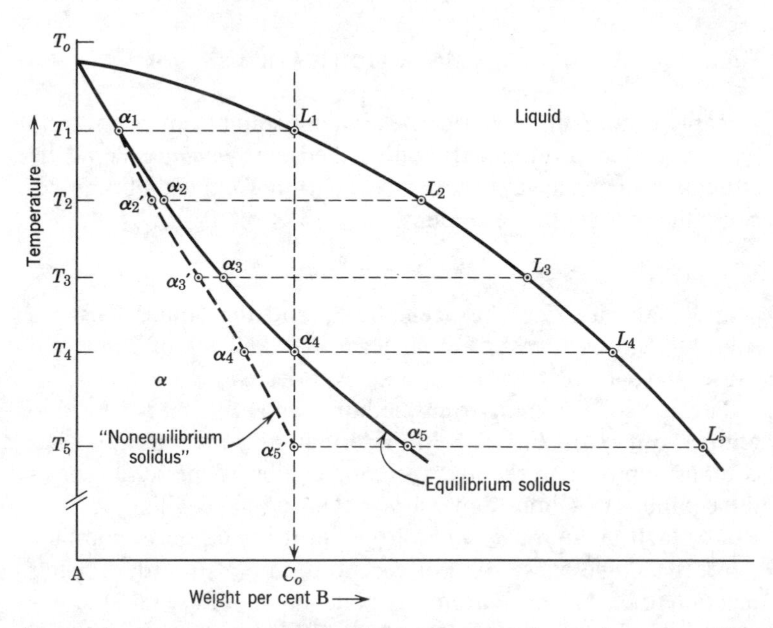

Figure 8.2 One way of considering the development of a cored structure. The alloy is not considered to be completely solid until its composition line crosses the "nonequilibrium solidus" at T_5.

A more quantitative picture of solidification under the conditions that result in coring may be obtained by writing a materials balance for the conservation of solute atoms during the solidification of an increment of material. At temperature T,

$$(L - \alpha)\, df_s = (1 - f_s)\, dL \tag{8.1}$$

where L is the equilibrium concentration of the liquid, α is the equilibrium concentration of the solid, and f_s is the fraction of material that is solid. Note that the incremental amount of solid formed is proportional to $(1 - f_s)$; in other words, the less liquid left, the less solid formed in the same increment. Therefore the nonequilibrium solidus of Figure 8.2 never really crosses C_o under the conditions which we have stipulated. (See Problem 8.1 for the solution of Equation 8.1.) An alloy that solidifies under these conditions will not become *completely* solid until the temperature is decreased to an invariant transformation temperature (eutectic, peritectic, etc. or the melting point of a pure component).

8.3 SURROUNDING DURING SOLIDIFICATION

Rapid cooling through a peritectic temperature causes a second type of nonequilibrium structure called *surrounding.* Peritectic structures almost always show this departure from equilibrium because the product of the reaction

$$L_p + \alpha_p \xrightarrow{\text{cooling}} \beta_p$$

forms at the interface between the α_p and the liquid. Being a solid phase, the β_p creates a barrier to the diffusion of atoms and causes the reaction to proceed at an ever-decreasing rate.

The first solid forming from the liquid alloy C_o as it is cooled to temperature T_1 in Figure 8.3*a* under equilibrium conditions has composition α_1, and the average composition of the solid follows the equilibrium solidus down to α_p. During the equilibrium peritectic reaction, all the α_p and all the peritectic liquid L_p combine to form a single phase solid of composition β_p. In order for this reaction to take place, B atoms from the liquid diffuse into the surface of the α_p and create a thin shell of β_p which *surrounds* the α_p phase (Figure 8.3*b*). More β_p is now produced at the interface be-

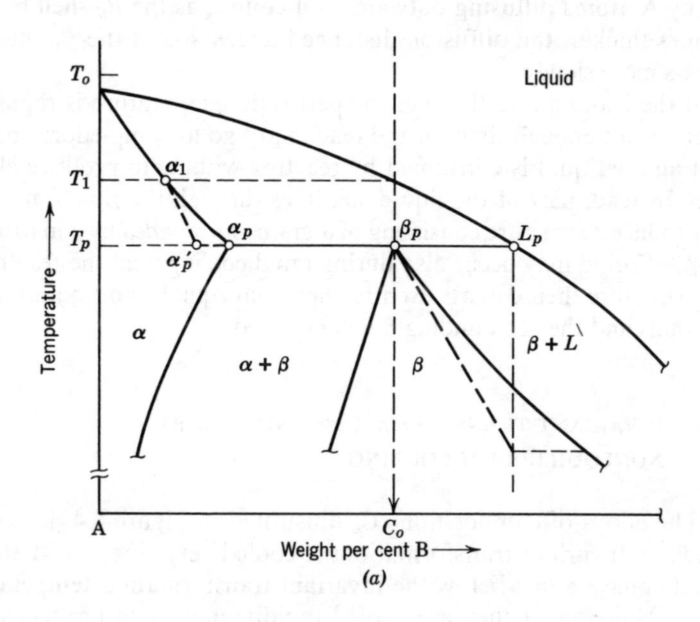

(a)

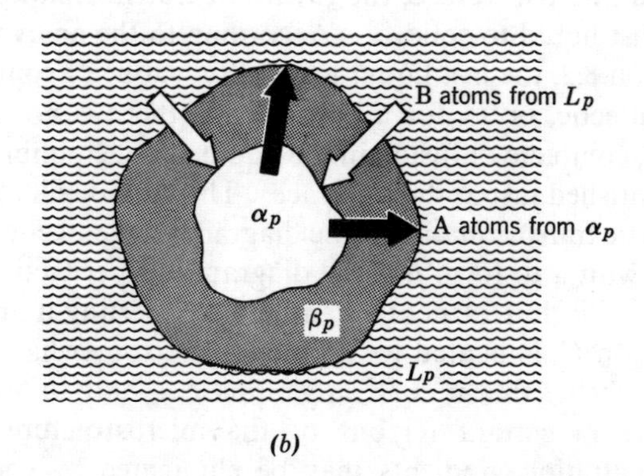

(b)

Figure 8.3 Surrounding during nonequilibrium cooling. *(a)* The peritectic portion of a phase diagram. If an alloy C_o is cooled rapidly, surrounding may result. *(b)* Atom transport during a peritectic reaction.

tween α_p and β_p by B atoms diffusing inward and between β_p and L_p by A atoms diffusing outward. Of course, as the β_p shell becomes thicker, the diffusion distance increases, and the β_p shell grows more slowly.

If the cooling rate through the peritectic temperature is rapid, there is not enough time for the reaction to go to completion, and not all the liquid is consumed by reacting with α_p to produce all β_p. Instead, part of the liquid solidifies through the $\beta + L$ field to produce a structure consisting of α grains *surrounded* by a matrix of β. Coring may occur also during rapid cooling, and the resulting structure then departs even further from equilibrium: both the α grains and the surrounding β will be cored.

8.4 INVARIANT TRANSFORMATIONS INDUCED BY NONEQUILIBRIUM COOLING

The alloys of compositions C_o illustrated in Figure 8.4 do not undergo invariant transformations if cooled very slowly and are single phase α just below the invariant transformation temperature. However, if they are cooled rapidly enough to produce a cored microstructure, the invariant transformation can occur in the last liquid to solidify. According to the analysis presented in Section 8.2, the last liquid to solidify will be of composition L_e for an eutectic, or L_p for a peritectic, if there is no diffusion in the solid, complete diffusion in the liquid and if equilibrium conditions are satisfied just at the interface. This liquid will then undergo an eutectic transformation if the diagram is an eutectic, or it will combine with α to form β if the diagram is a peritectic. This type of nonequilibrium structure can be disconcerting if an alloy of composition C_o is heated to a temperature somewhat above T_e or T_p in the belief that it is all single-phase α, only to find that melting occurs in certain regions of the microstructure. In practice, concentration gradients may be eliminated by cooling the alloy extremely slowly. They are eliminated much more quickly, however, by rolling or forging a cored alloy to break up the structure mechanically, then homogenizing it at a temperature just under T_e or T_p. The strain induced by the deformation promotes very rapid atomic readjustments at the homogenizing temperature.

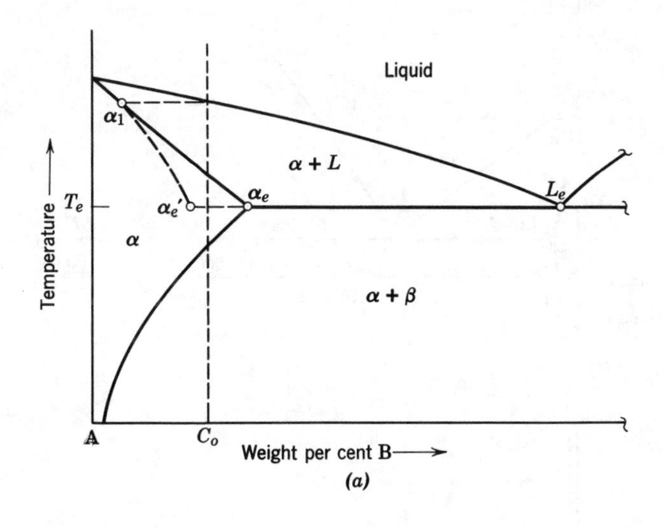

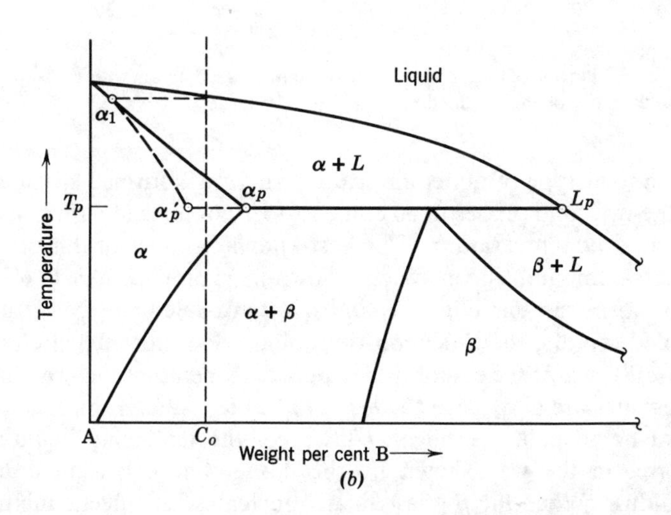

Figure 8.4 Nonequilibrium eutectic (*a*) and nonequilibrium peritectic reactions. In both cases, equilibrium cooling of alloy C_o would produce single phase α just below T_e or T_p. However, if the α is cored, some liquid exists at the invariant temperature and goes through the invariant reaction.

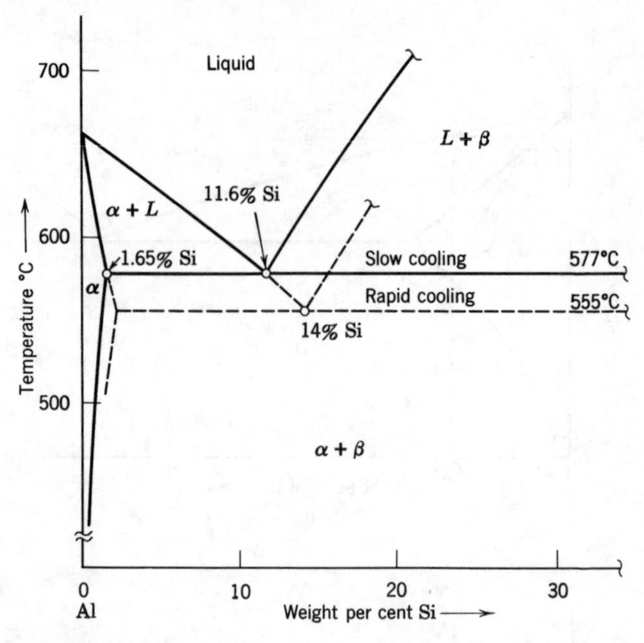

Figure 8.5 Portion of the aluminum-silicon equilibrium diagram with a superposed phase diagram for the modified nonequilibrium eutectic.

Another type of nonequilibrium structure is formed when one of the two solid phases in an eutectic does not nucleate until below the eutectic temperature. The best-known instance of this occurs in the aluminum-silicon system, shown in Figure 8.5. An 11.6% Si alloy forms an eutectic of α and β when cooled slowly. If it is cooled rapidly, the β (silicon-rich) phase does not nucleate until some 20° below the equilibrium eutectic temperature. The α phase does nucleate and, since the α phase has less silicon than the original liquid had, the silicon content of the remaining liquid increases in the way shown by the dashed-line extension of the liquidus. When the β phase finally nucleates, an eutectic mixture results in which there is more β and less α than in the equilibrium eutectic. Therefore a rapidly cooled 11.6% Si alloy will have a microstructure containing proeutectic α and the nonequilibrium eutectic, and a 14% Si alloy will have a completely eutectic microstructure.

8.5 SUPERSATURATION AND PRECIPITATION

The rate of atomic diffusion varies as $e^{-C/T}$, often written $\exp(-C/T)$, where C is approximately a constant and T is absolute temperature. Therefore phase transformations which occur by diffusion often can be completely, or almost completely, inhibited by cooling the solid rapidly to a low temperature. Practical advantage is taken of this fact in so-called *precipitation-hardening* alloys. The reason for precipitation in such an alloy is illustrated in the phase diagram in Figure 8.6. An alloy, C_o, may be *solutionized* (put into a condition where all the B atoms are in solid solution α) by holding it at a temperature T_o. In practice, T_o is specified a little below the eutectic temperature in the event that the alloy is severely cored. If the solutionized alloy is cooled from T_o, it remains single phase down to a temperature T_1, where the over-all composition line crosses the solvus (the limit of solubility of B in α). On further slow cooling to T_2, the maximum equilibrium solubility of B in α is decreased to α_2, and some B atoms

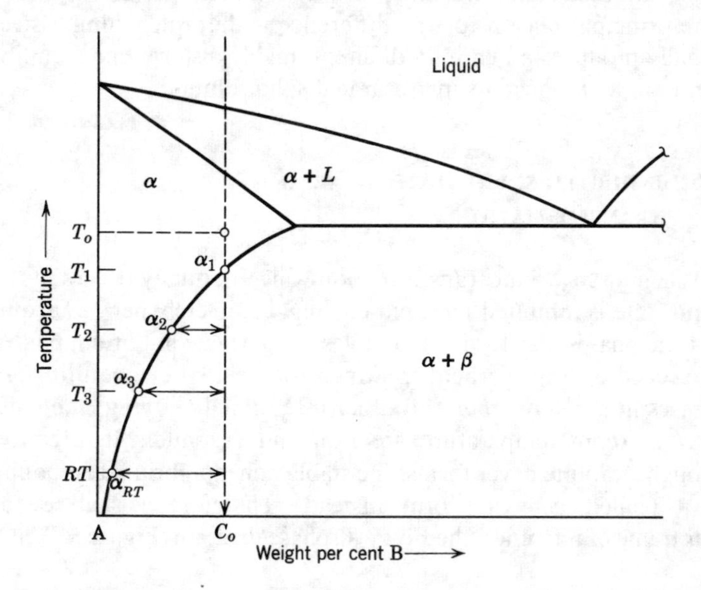

Figure 8.6 Precipitation from solid solution during cooling from the α region into the $\alpha + \beta$ region.

are rejected in the form of a β precipitate. On lowering the temperature to T_3, more β (the amount may be determined by applying the lever rule) is precipitated to maintain equilibrium.

The way in which the precipitation process is controlled in order to increase the strength of an alloy is as follows. The alloy is solutionized at temperature T_o; it is then *quenched* to some low temperature, say room temperature, at which the alloy will retain all the excess B atoms in a nonequilibrium α solid solution. This nonequilibrium solid solution is called *supersaturated.* If the alloy is then heated to a temperature where the nucleation rate is high but the growth rate is low, a large number of submicroscopic precipitate particles with a nonequilibrium transition structure may form. These particles strain the matrix (the α solid solution) so much that they increase its strength and hardness. In precipitation-hardenable alloys, the maximum hardness is always associated with the early stages of precipitation, and not with the quenched, super-saturated structure. In the quenching and tempering of steel, discussed next, the reverse is true; maximum hardness occurs in the quenched, supersaturated condition, and precipitation decreases the hardness. The principal reason for the difference is that quenching a steel usually produces an entirely different, highly distorted nonequilibrium phase, not just a supersaturated solid solution.

8.6 INHIBITED SOLID STATE INVARIANT TRANSFORMATIONS

When an invariant transformation which normally occurs in the solid state is inhibited by rapid cooling, completely new, nonequilibrium phases may form. Examples are the nonequilibrium phases produced during the heat treatment of steel.[1] The equilibrium phases in a *plain carbon* (no other intentional alloying elements) *steel* at room temperature are iron and graphite. In practice, though, graphite never forms; the stable nonequilibrium compound Fe_3C (called *cementite*) forms instead. Therefore we shall use, as our frame of reference, the Fe-Fe_3C phase diagram (Figure 8.7). Of

[1] Steel is primarily an alloy of carbon in iron but often contains other alloying elements as well.

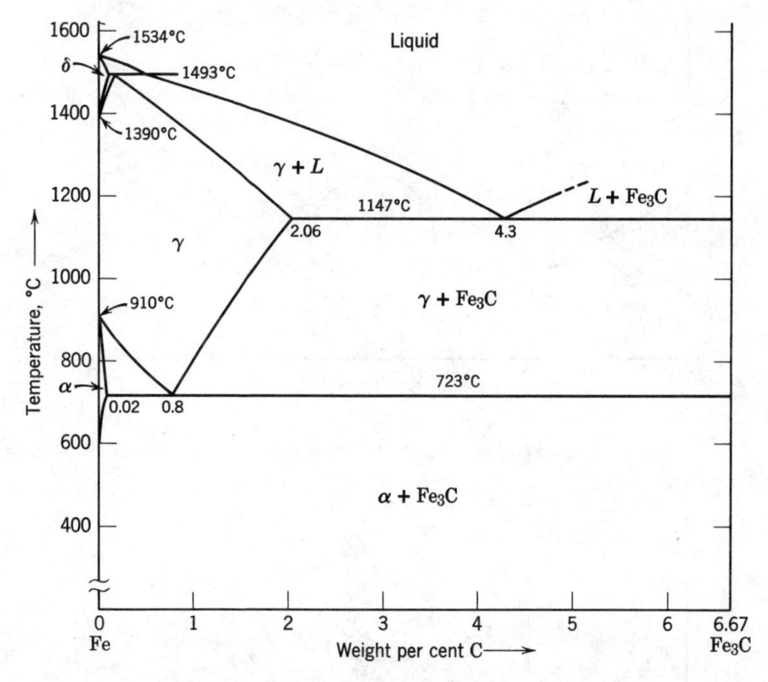

Figure 8.7 Phase diagram for the system Fe-Fe_3C.

the three invariant transformations (peritectic, eutectic, and eutectoid) shown in this diagram, the only one which takes place completely in the solid state is the eutectoid. Because solid state diffusion is relatively slow, this transformation can be inhibited completely by quenching the steel rapidly enough from a temperature above 723°C.

In order to understand the origin of the nonequilibrium structures which result from inhibiting the eutectoid decomposition of steel, it is necessary first to discuss the structures of slowly cooled, plain carbon steels in which the transformation occurs. Carbon is much more soluble in the FCC phase (labeled γ and called *austenite*) than it is in the BCC phases (α, called *ferrite,* and δ) because the interstitial holes between FCC packed atoms are larger than those between BCC packed atoms. Therefore an alloy containing more than 0.02% C will, on slow cooling from the austenite region, precipitate the excess carbon in the form of Fe_3C. The pro-

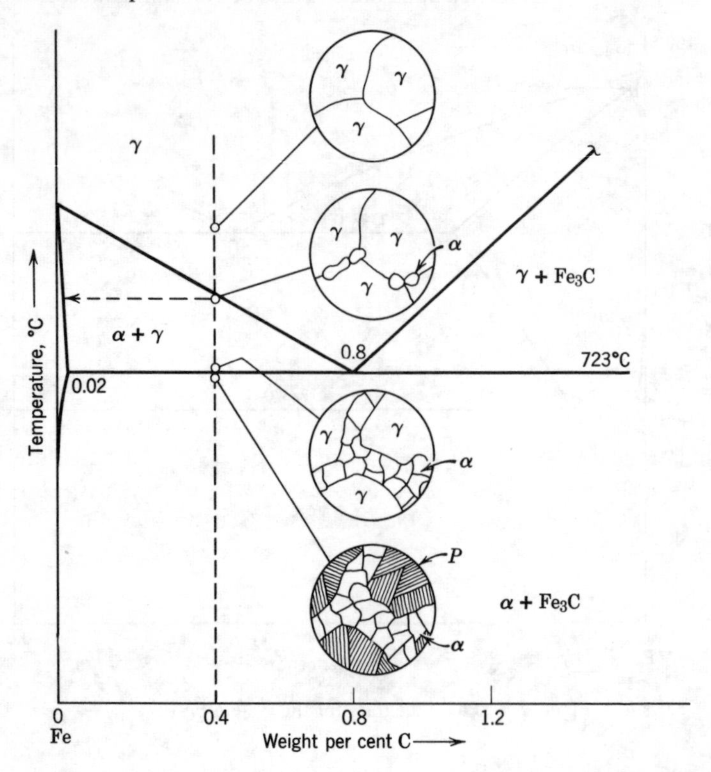

Figure 8.8 A schematic representation of the microstructural changes which occur during the slow cooling of a 0.4% C steel. (*a*) Austenite (γ). (*b*) Formation of α grains at γ grain boundaries. (*c*) Growth of α at grain boundaries; composition of γ is now 0.8% C. (*d*) 0.8% C γ transforms to pearlite during cooling below 723°C.

gressive changes in the microstructure of a slowly cooled 0.4% C steel are illustrated schematically in Figure 8.8. First, grains of α solid solution (ferrite) nucleate at the austenite grain boundaries and grow until, at a temperature just above 723°C, the structure consists of γ (0.8% C) and α (0.02% C). At this temperature, the weight fraction of austenite is determined by the lever rule.

$$f_\gamma = \frac{0.4 - 0.02}{0.8 - 0.02} = 0.49$$

When the alloy is cooled below 723 °C, this austenite transforms to a fine, lamellar mixture of α and Fe_3C. The mixture is distinguished easily in the microscope (see Figure 6.9) and is called *pearlite.* It is the mixture already described in Chapter 6. The behavior of a 1% C steel is similar except that there is a larger fraction of austenite just above 723 °C,

$$f_\gamma = \frac{6.67 - 1.0}{6.67 - 0.8} = 0.97$$

and the proeutectoid phase is Fe_3C rather than α. Microstructures like these, in which the observed low temperature phases are ferrite and cementite, are said to result from *normalizing*, or slow cooling, the steel. The compositions of the phases and their relative amounts can be determined from the Fe-Fe_3C phase diagram.

If the steel is quenched instead of normalized, two microstructures other than pearlite may be produced by inhibiting the eutectoid transformation.

1. *Bainite* may form when austenite is quenched to some temperature, typically in the range between 200 and 400 °C, and held there. Bainite is a dispersion of submicroscopic carbides in a highly strained α matrix that contains more than 0.02% C.
2. *Martensite* may form when austenite is quenched to lower temperatures than those at which bainite will form. It is an extremely hard and brittle phase in which all the carbon is trapped in supersaturated solid solution. The excess carbon distorts the crystal structure to body-centered tetragonal (BCT), the amount of distortion (measured by the average c/a ratio of the tetragonal unit cell) being roughly proportional to the carbon content.

The transformation of austenite to pearlite and bainite, as well as to proeutectoid ferrite and cementite, appears to occur by a process of nucleation and growth and is thus controlled by diffusion rate. These transformations are both time dependent and temperature dependent. The transformation from austenite to martensite, on the other hand, is diffusionless and occurs so rapidly that it is almost independent of time. It occurs by a shear mechanism similar to that responsible for mechanical twinning (see Figure 4.14), and the fraction of austenite that transforms to martensite is determined almost completely by the temperature; the lower the temperature, the larger the fraction transformed until it is all martensite.

The rate at which the equilibrium decomposition of austenite occurs is governed in a plain carbon steel by how fast the carbon diffuses through the austenite, since α will not form in the microstructure until the local concentration of carbon is about 0.02%, and cementite will not form until the local concentration of carbon is about 6.67%. However, in an alloy steel, not only the carbon has to redistribute itself but so do the alloying elements. Alloying elements like chromium, manganese and molybdenum require considerable redistribution, but they diffuse so slowly that the transformation from austenite to pearlite and ferrite, or to pearlite and cementite, is very sluggish. Since the equilibrium reaction takes so much more time to start in one of these alloy steels, martensite can be formed at much less drastic cooling rates. We say that this alloy is easy to harden, or that it has a higher *hardenability.* In fact, a thin piece of steel containing a few per cent of chromium and molybdenum will transform completely to martensite even if it is merely cooled in air and not quenched. Alloying elements may also retard the equilibrium transformation by depressing the eutectoid temperature to a region where all diffusion is much slower. If enough nickel is added to steel, the eutectoid temperature is lowered so greatly that, when it is reached, the transformation cannot occur and the steel remains *austenitic.*

8.7 ISOTHERMAL TRANSFORMATION DIAGRAMS

The usual type of phase diagram does not represent the course which the time-dependent decomposition of a phase follows. This process may be graphed, as in Figure 8.9 for austenite decomposition, by plotting per cent transformed as a function of time at a particular temperature. The data for such a graph are determined by solutionizing (called *austenitizing* in this special case) a number of thin samples of a given steel and then quenching them to some temperature below the eutectoid temperature. Samples are periodically removed after *isothermal* (constant temperature) decomposition starts to take place and are quenched to room temperature or below. The extent of transformation prior to the last quench may be determined by examining metallographic sections of the samples. The samples shown schematically in Figure 8.9 were quenched initially to a temperature high enough that the austenite transformed to pearlite; at some lower temperature it would have transformed to bainite.

If the process above is repeated, and the alloy is quenched each time to a different isothermal transformation temperature, a family of curves can be obtained which are cross-plotted, as in Figure 8.10, to yield curves of per cent austenite transformed as a function of temperature and time. These curves are called *T-T-T curves* (time-temperature-transformation).

A *T-T-T* curve for an eutectoid (0.8% C) plain carbon steel is shown in Figure 8.11; the curves labeled "start" and "finish" describe the start and finish of the austenite decomposition. The microstructures that result from the decomposition are determined by the temperature. At temperatures just below the eutectoid temperature, the diffusion rate is high, and the nucleation rate is low; therefore few nuclei form, but they grow rapidly and produce a coarse pearlitic structure with widely spaced lamellae. At temperatures a couple of hundred degrees C lower, the nucleation rate is much higher, and diffusion is slower, so a finer pearlite results. A few hundred degrees below this, and bainite is the product of the transformation. At still lower temperatures, where the cooling curve intersects the M_s (the start of the martensite transformation), martensite forms. If the cooling curve does not cross the M_f (the finish of the martensite transformation), some austenite is retained.

The *T-T-T* curve exhibits a "nose," or minimum time, before which the transformation does not begin. Alloy steels often show a second, lower nose (sometimes called a "chin") in the range of temperatures where bainite is formed. If martensite is to be produced, the steel must be quenched rapidly enough that neither the "nose" nor the "chin" is crossed. The minimum time interval before the start of transformation (called the "gate") dictates the severity of the quench necessary. The purpose of adding hardenability-improving alloying elements is to widen this gate and thereby make it possible to produce martensite in thick sections without having to quench it so rapidly that surface cracks form.

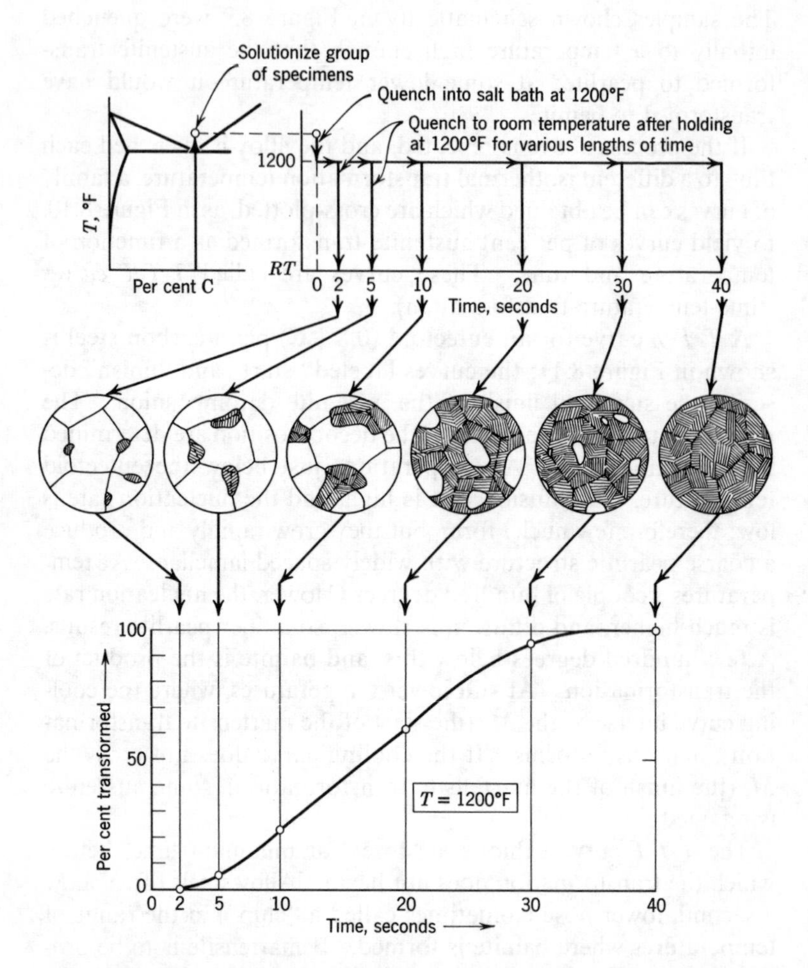

Figure 8.9 An experimental method for determining a curve of extent of isothermal transformation of a eutectoid steel versus time. In this example, the transformation temperature (1200°F) is one at which austenite decomposes to pearlite. In the schematic microstructures, cross-hatched regions represent pearlite and light regions represent other lower temperature transformation products.

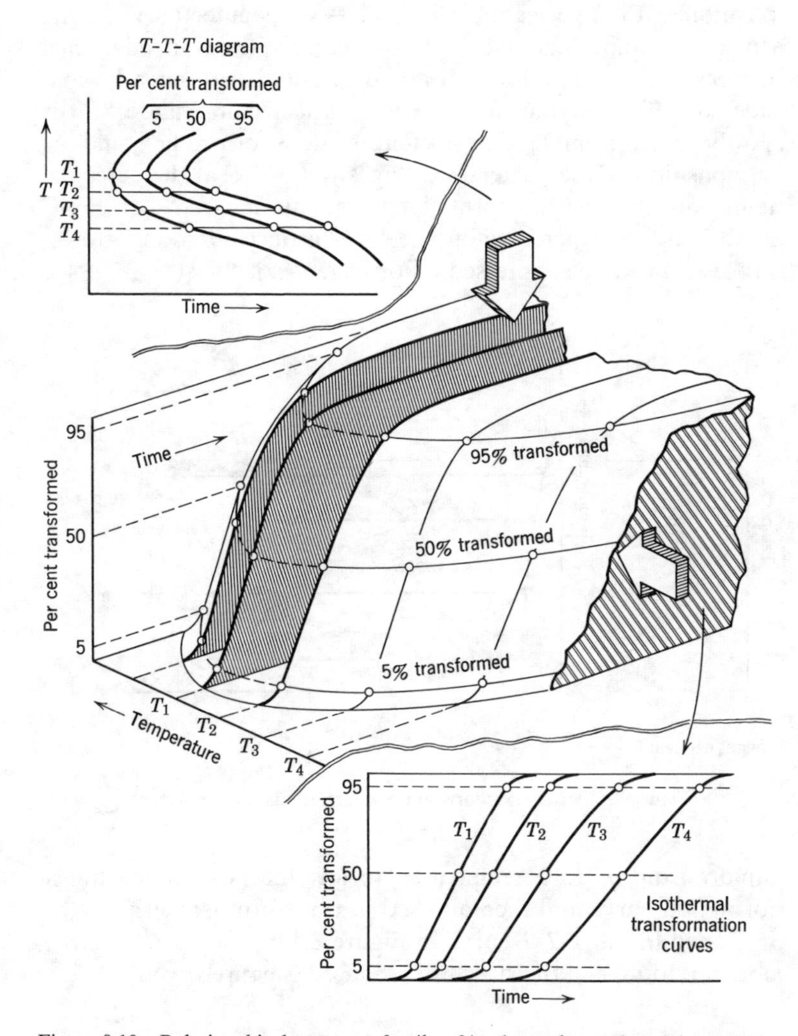

Figure 8.10 Relationship between a family of isothermal transformation curves and the T-T-T diagram produced from them. Both are sections through the three-dimensional time-temperature-transformation surface shown in the center of this figure.

The *T-T-T* curve in Figure 8.11 was for a steel of eutectoid composition. If the per cent carbon is less (hypoeutectoid) or more (hypereutectoid) than the eutectoid composition, an additional factor—the precipitation of a proeutectoid phase—must be considered. If the hypoeutectoid steel indicated in Figure 8.12 is slowly cooled from T_o, proeutectoid ferrite is formed first, and the composition of the austenite changes to 0.8% C at the eutectoid temperature. Since the formation of proeutectoid ferrite (or Fe_3C in the case of hypereutectoid steels) is a nucleation and growth process, it can be suppressed. For instance, if the steel is cooled

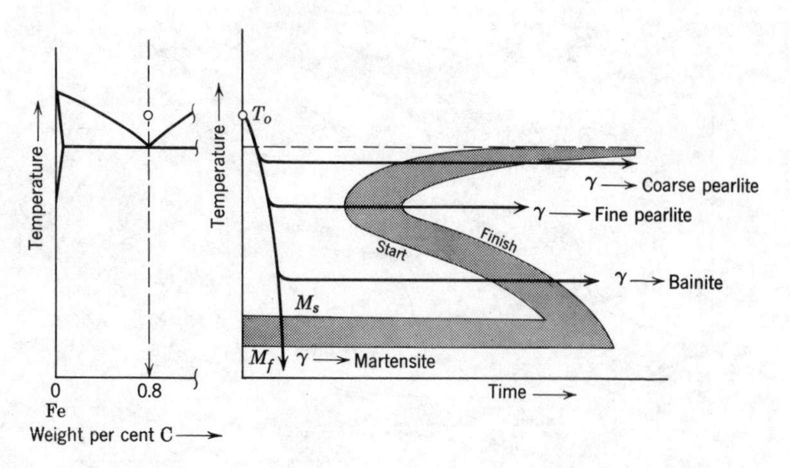

Figure 8.11 A *T-T-T* curve for a eutectoid plain carbon steel.

rapidly from T_o to a temperature below the nose of the transformation curve and then allowed to transform isothermally, as indicated in the *T-T-T* curve in Figure 8.12, proeutectoid ferrite does not form, and the structure produced is entirely bainitic.

8.8 NONEQUILIBRIUM TRANSFORMATIONS AND THE HEAT TREATMENT OF ALLOYS

The best examples of changes which may be produced in the structure of a material by thermal treatment are found in alloy systems, since in materials other than metals interatomic bonding is so strong that diffusion is very slow at all temperatures except near the solidus. The reasons for the heat treatments given to different alloys can be understood in terms of the effects of time and temperature on their structures.

Castings exhibiting either coring or surrounding result from cooling the alloy too quickly for equilibrium to be maintained by diffusion. Such castings often are heat treated to *homogenize* the structure, or bring it closer to equilibrium. They are reheated to an elevated temperature, which promotes diffusion to bring about a

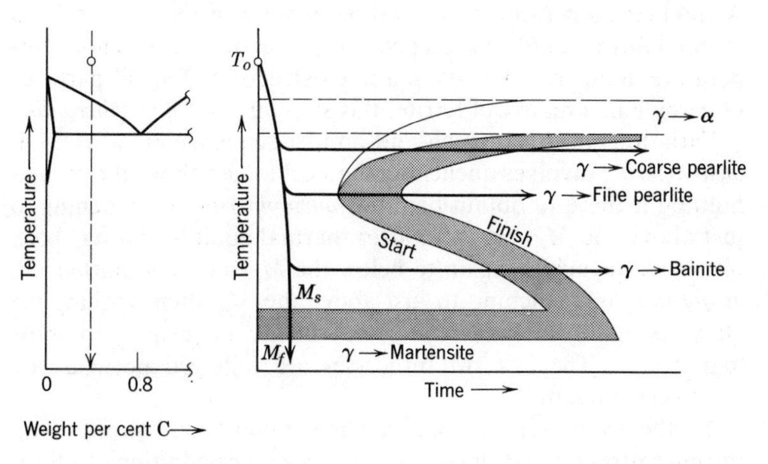

Figure 8.12 A *T-T-T* curve for a hypoeutectoid plain carbon steel.

more equilibrium structure, but which does not melt any nonequilibrium phases. If the unhomogenized material is ductile, the homogenizing treatment may be shortened considerably by rolling or forging the casting first, since atomic rearrangements occur more readily in a strained material.

Precipitation-hardenable alloys may also be cored in the as-cast condition. Such alloys are usually hot-worked, then homogenized and solutionized at the same temperature, and finally quenched. Precipitation may be induced by reheating the alloy to a slightly elevated temperature, during which time it is said to *precipitation harden.* In some alloys precipitation occurs naturally at room temperature; these alloys are said to *age harden.* In either case, if the precipitation proceeds too far, the alloy becomes softer again; it is then said to have *overaged.*

Unlike precipitation-hardenable alloys which remain soft after quenching and harden when precipitation begins, steel is hardened by quenching and is softened by precipitation. The reason for the difference is the allotropic transformation in steel which leads to the formation of martensite. Martensite can be softened by precipitating some of the carbon as a carbide; the process is called *tempering.* Low-temperature tempering produces a very finely dispersed two-phase structure of carbide in low-carbon martensite. At higher temperatures, the carbon content of the martensite is reduced further until, at a temperature just under the eutectoid temperature, tempering results in a microstructure of small particles of carbide in a matrix of ferrite; this structure is called *spheroidite.*

Variations of this quench-and-temper sequence are also used: *austempering* involves quenching the steel to just above the M_s and holding it there to obtain bainite; *martempering* is quenching to just above the M_s and then, after thermal equilibrium has been obtained, quenching again to below the M_f to obtain martensite; *ausforming* is quenching to just above the M_s, then working the alloy to distort the austenite, and finally quenching to form martensite. The last two processes are followed normally by a tempering treatment.

Another example of the manipulation of properties by controlling the microstructure occurs in the *cast irons.* The addition of silicon and more carbon to steel results in the formation of graphite as the carbon-rich phase when the alloy is cooled slowly. These Fe-Si-C alloys are called cast irons. If graphite is the only carbon-rich phase formed, the phases obtained are those described by the dashed lines in the Fe-C diagram in Figure 8.13. On the other hand, Fe_3C can be made to form by cooling the alloy more rapidly. If the cooling rates are such that graphite forms only at higher temperatures, the iron-carbon diagram and the iron-iron carbide diagram must be used in conjunction to rationalize the resulting structures.

Cooling a cast iron to just above the eutectic temperature causes the formation of primary austenite (proeutectic phase) dendrites. If the alloy is cooled rapidly through the eutectic temperature,

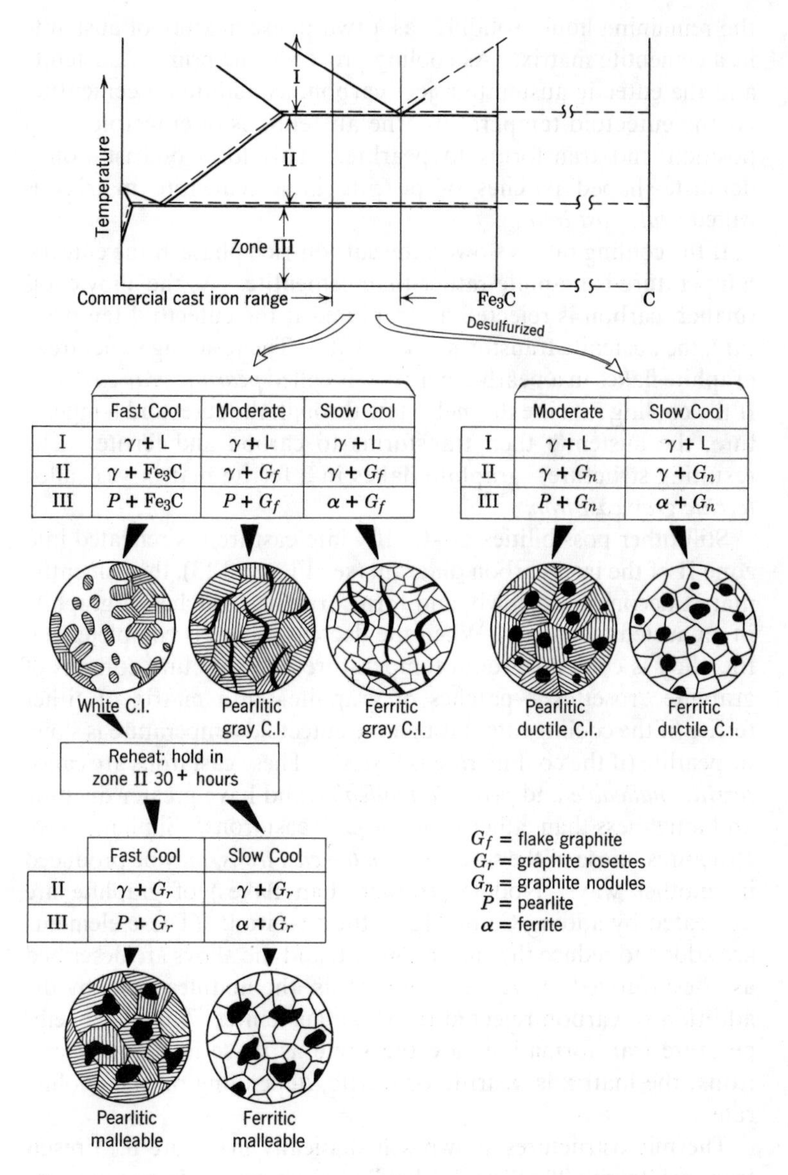

Figure 8.13 Cast iron microstructures and a summary of the phases coexisting at various temperatures.

the remaining liquid solidifies as a two-phase mixture of austenite in a cementite matrix. As cooling proceeds, the primary austenite and the eutectic austenite reject carbon, as additional cementite. At the eutectoid temperature, the austenite is of eutectoid composition and transforms to pearlite. This form of cast iron—dendrite-shaped patches of pearlite in a cementite matrix—is called *white cast iron.*

If the cooling rate is slower, the carbon-rich phase at the eutectic temperature is graphite rather than cementite. As the alloy cools further, carbon is rejected as graphite; at the eutectoid temperature, the austenite transforms to pearlite. The resulting structure—graphite flakes in a pearlite matrix—is called *pearlitic gray cast iron.* If the cooling rate is extremely slow through the eutectoid temperature, the austenite then transforms to carbon and ferrite. The resulting structure—graphite flakes in a ferrite matrix—is called *ferritic gray cast iron.*

Still other possibilities exist. If white cast iron is reheated into zone II of the iron-carbon diagram (see Figure 8.13), the cementite matrix decomposes slowly and forms irregular patches of graphite in an austenite matrix. When the decomposition is complete, and the alloy is cooled to room temperature, the structure consists of graphite "rosettes"—patches of graphite—in a matrix of either ferrite (if the cooling rate through the eutectoid temperature is slow) or pearlite (if the cooling rate is faster). These cast irons are called *ferritic malleable* and *pearlitic malleable* and have greater ductility and toughness than either white or gray cast irons. Similar microstructures, those of the so-called *ductile cast irons,* may be produced in another way. Nodules (rather than flakes) of graphite are nucleated by adding Ce or Mg to the cast iron. (These elements are added to reduce the sulfur content, and the alloys are described as "desulfurized".) As the alloy cools, the nodules grow by the addition of carbon rejected from the austenite. The lower temperature transformations are the same as those in the gray cast irons: the matrix is pearlite or ferrite, depending on the cooling rate.

The microstructures shown schematically in Figure 8.13 result from arbitrarily "fast" and "slow" cooling rates. In practice, intermediate cooling rates usually prevail and result in hybrid microstructures. One example occurs in the gray cast irons cooled at a

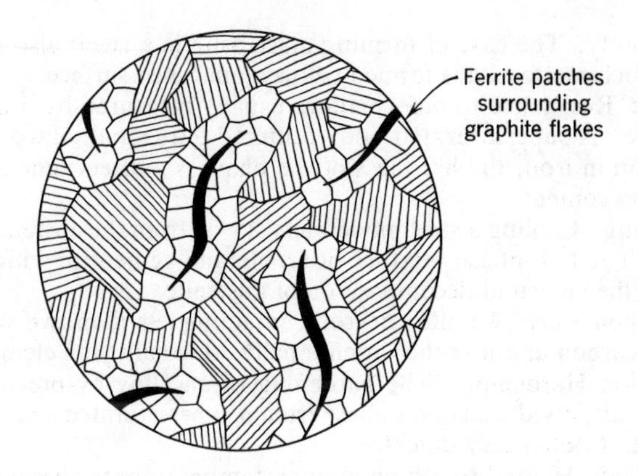

Figure 8.14 Schematic microstructure of commercial pearlitic gray cast iron. Partial decomposition of pearlite adjacent to the graphite flakes produces ferrite regions surrounding the graphite flakes. Compare this representation with Figure 6.10.

rate between "moderate" and "slow". The pearlite is only partially decomposed, and the resulting structure is a pearlite matrix containing graphite flakes surrounded by ferrite. The microstructure is shown schematically in Figure 8.14 and actually in Figure 6.10, already discussed. Similar hybrid structures are found in ductile and malleable cast irons.

DEFINITIONS

Austenite: A face-centered cubic, iron-rich solid solution containing carbon (and possibly other elements).

Austenitizing: Heating a steel to a temperature where austenite is the stable phase.

Bainite: A fine dispersion of iron carbide in a strained ferrite matrix; formed by isothermal decomposition of austenite.

Cast Irons: Iron-carbon alloys which contain more than about two weight per cent carbon.

Cementite: The compound, Fe_3C.

Coring: A nonequilibrium microstructural feature in which concentration gradients result from cooling an alloy too rapidly through a two phase field.

Ferrite: A body-centered cubic (alpha), iron-rich solid solution containing carbon (and possibly other elements).

Hardenability: The ease of forming martensite in a steel; also the depth to which martensite is formed below a quenched surface.
Hardness: Resistance to deformation, usually measured by indentation.
Martensite: A supersaturated body-centered tetragonal solid solution of carbon in iron; the hardness of the phase is a direct function of the carbon content.
Normalizing: Cooling a steel moderately slowly from the austenite region.
Pearlite: The two phase mixture of ferrite and cementite which results from the eutectoid decomposition of austenite.
Plain Carbon Steel: An alloy of iron containing less than two weight per cent carbon and no other intentionally added alloying elements.
Precipitation Hardening: The hardening of an alloy by precipitating a finely dispersed transition phase from a supersaturated solid solution.
Quenched: Cooled very quickly.
Solutionized: Heated to a high enough temperature to dissolve second phase particles, resulting in a single phase solid.
Spheroidite: A mixture of microscopically resolvable particles of cementite in a matrix of ferrite.
Supersaturated Solid Solution: A solid solution containing more than the equilibrium amount of solute.
Surrounding: A nonequilibrium microstructural feature in which the phase formed by peritectic decomposition surrounds one of the high temperature phases, inhibiting further transformation.
T-T-T Curves: Time-temperature-transformation curves, in which temperature is plotted versus time for various degrees of completion of a phase transformation; the curves are determined experimentally under isothermal conditions.
Tempering: Heating a martensitic steel to precipitate some of the carbon as a carbide, increasing both the softness and the toughness of the material.

BIBLIOGRAPHY

INTRODUCTORY REFERENCE:

V. F. Zackay, "The Strength of Steel," *Scientific American,* Vol. 209 (August, 1963) p. 72.

SUPPLEMENTARY REFERENCES:

R. M. Brick and A. Phillips, *Structure and Properties of Alloys,* McGraw-Hill Book Co., N. Y. (1949).

T. A. Read, "Phase Transformations in Metals and Their Influence on Mechanical Properties," *The Science of Engineering Materials,* ed. by J. E. Goldman, John Wiley and Sons, N. Y. (1957) p. 196.

Principles of Heat Treatment, American Society for Metals, Metals Park (1964).

MORE ADVANCED TEXT:

Phase Transformations in Solids, ed. by R. Smoluchowski, J. E. Mayer and W. A. Weyl, John Wiley and Sons, N. Y. (1951).

PROBLEMS

8.1 Solve Equation 8.1 by assuming that α/L = constant and by integrating from completely liquid to some arbitrary fraction solidified.

8.2 The Ge-Si equilibrium diagram, like that of Cu-Ni, shows a complete range of solid solubility. Explain why Si-Ge alloys are much more difficult to homogenize than Cu-Ni alloys.

8.3 The K_2O-SiO_2 equilibrium diagram contains two intermediate compounds between 56 and 100 wt.% silica: $K_2O \cdot 2SiO_2$ ($T_m = 1050°C$) and $K_2O \cdot 4SiO_2$ ($T_m = 790°C$): eutectics exist at approximately 68% SiO_2 and 750°C and at 75% SiO_2 and 770°C.

(a) Sketch the equilibrium diagram and show how it is probably changed by the fact that $K_2O \cdot 4SiO_2$ does not nucleate normally.

(b) In the case where $K_2O \cdot 4SiO_2$ does not nucleate, a eutectic still is observed; give its estimated composition and temperature.

8.4 On a sketch of the Ag-Cu equilibrium diagram (Appendix VI) show the series of treatments required to transform a cored casting to a wrought precipitation-hardened part.

8.5 Aluminum alloy rivets are often of a composition such that they can be deformed easily when put in place but age-harden at normal temperatures. How could one salvage a batch of rivets which had age-hardened before they had been used?

8.6 What is retained austenite? Describe a procedure for eliminating its presence in a piece of hardened steel.

8.7 What is the reason for the characteristic "S" shape of the per cent transformation versus time curves in Figures 8.9 and 8.10?

8.8 Draw cooling curves for (a) quenching and tempering, (b) austempering, (c) martempering, and (d) ausforming, superimposed on the *T-T-T* curve for a hypereutectoid steel. Are all these processes isothermal?

8.9 Explain qualitatively why increasing the carbon content of a steel from 0.02 to 0.8% lowers the M_s and M_f and also widens the "gate" of the *T-T-T* curve.

8.10 The hardness of martensite depends primarily on the carbon content and is little influenced by the presence of alloying elements in the steel. Why are alloying elements added to steel?

8.11 Distinguish between the hardness and hardenability of a steel.

Give an approximate composition of a high hardness, low hardenability steel. What would you add to increase the hardenability?

8.12 What would you expect the effect of austenite grain size to be on the hardenability of a steel? Why?

8.13 Assume, as a rough approximation, that a continuous cooling curve can be approximated by small isothermal segments on a *T-T-T* diagram. Sketch cooling curves for the surface and for the center of a steel part which, on quenching, shows a martensitic surface and a pearlitic center. If the cooling curve for the surface represents the most drastic quench which can be attained, what modification of the steel can be made, without changing the surface hardness, to obtain a part hardened completely through its cross-section?

8.14 As in Problem 8.13, sketch cooling curves for the surface and center of an alloy steel part which is to be fully hardened by martempering. What is the advantage of this process compared to continuous cooling to room temperature?

8.15 How must a *T-T-T* curve be modified for the cases where phase transformations occur during continuous cooling rather than isothermally? Can bainite be formed in a plain carbon steel that is cooled at a constant rate (degrees/minute)?

8.16 Explain why the graphite phase in the Fe-C eutectic is much less continuous than the Fe_3C phase in pearlite.

8.17 Suggest a reason for the observation that gray cast iron is much more machinable than white cast iron.

8.18 Rank the following cast irons in order of increasing hardness: white cast iron, ferritic gray cast iron, pearlitic gray cast iron, "commerical" pearlitic gray cast iron.

Appendix IA Periodic Table of the Elements.

GROUP

PERIOD	IA	IIA	IIIB	IVB	VB	VIB	VIIB	VIII			IB	IIB	IIIA	IVA	VA	VIA	VIIA	0
1	1 H																	2 He
2	3 Li	4 Be											5 B	6 C	7 N	8 O	9 F	10 Ne
3	11 Na	12 Mg											13 Al	14 Si	15 P	16 S	17 Cl	18 Ar
4	19 K	20 Ca	21 Sc	22 Ti	23 V	24 Cr	25 Mn	26 Fe	27 Co	28 Ni	29 Cu	30 Zn	31 Ga	32 Ge	33 As	34 Se	35 Br	36 Kr
5	37 Rb	38 Sr	39 Y	40 Zr	41 Nb	42 Mo	43 Tc	44 Ru	45 Rh	46 Pd	47 Ag	48 Cd	49 In	50 Sn	51 Sb	52 Te	53 I	54 Xe
6	55 Cs	56 Ba	57 La	72 Hf	73 Ta	74 W	75 Re	76 Os	77 Ir	78 Pt	79 Au	80 Hg	81 Tl	82 Pb	83 Bi	84 Po	85 At	86 Rn
7	87 Fr	88 Ra	89 Ac															

Lanthanons

58 Ce	59 Pr	60 Nd	61 Pm	62 Sm	63 Eu	64 Gd	65 Tb	66 Dy	67 Ho	68 Er	69 Tm	70 Yb	71 Lu

Actinons

90 Th	91 Pa	92 U	93 Np	94 Pu	95 Am	96 Cm	97 Bk	98 Cf	99 Es	100 Fm	101 Md	102 No	103 Lw

Appendix IB Electronic Configuration of the Elements.

Atomic Number	Element	Number of Electrons in Each Group									
		$1s$	$2s$	$2p$	$3s$	$3p$	$3d$	$4s$	$4p$	$4d$	$4f$
1	H	1									
2	He	2									
3	Li	Helium core	1								
4	Be		2								
5	B		2	1							
6	C		2	2							
7	N		2	3							
8	O		2	4							
9	F		2	5							
10	Ne		2	6							
11	Na		Neon core		1						
12	Mg				2						
13	Al				2	1					
14	Si				2	2					
15	P				2	3					
16	S				2	4					
17	Cl				2	5					
18	Ar				2	6					
19	K			Argon core				1			
20	Ca							2			
21	Sc						1	2			
22	Ti						2	2			
23	V						3	2			
24	Cr						5	1			
25	Mn						5	2			
26	Fe						6	2			
27	Co						7	2			
28	Ni						8	2			
29	Cu						10	1			
30	Zn						3d Filled	2			
31	Ga							2	1		
32	Ge							2	2		
33	As							2	3		
34	Se							2	4		
35	Br							2	5		
36	Kr							2	6		

Atomic Number	Element	$1s$ to $4p$ Filled	Number of Electrons in Each Group							
			$4d$	$4f$	$5s$	$5p$	$5d$	$5f$	$5g$	$6s$
37	Rb	Krypton core			1					
38	Sr				2					
39	Y		1		2					
40	Zr		2		2					
41	Nb		4		1					
42	Mo		5		1					
43	Tc		6		1					
44	Ru		7		1					
45	Rh		8		1					
46	Pd		10							
47	Ag		$4d$ Filled		1					
48	Cd				2					
49	In				2	1				
50	Sn				2	2				
51	Sb				2	3				
52	Te				2	4				
53	I				2	5				
54	Xe			0	2	6				
55	Cs	$1s$ to $4d$ Filled		0	$5s$ and $5p$ Filled					1
56	Ba			0						2
57	La			0			1			$6s$ Filled
58	Ce			2			0			
59	Pr			3			0			
60	Nd			4			0			
61	Pm			5			0			
62	Sa			6			0			
63	Eu			7			0			
64	Gd			7			1			
65	Tb			8			1			
66	Dy			10			0			
67	Ho			11			0			
68	Er			12			0			
69	Tm			13			0			
70	Yb			14			0			
71	Lu			14			1			
72	Hf			14			2			

Atomic Number	Element	$1s$ to $5p$ Filled	Number of Electrons in Each Group						
			$5d$	$5f$	$5g$	$6s$	$6p$	$6d$	$7s$
73	Ta	$1s$ to $5p$ Filled	3			2			
74	W		4			2			
75	Re		5			2			
76	Os		6			2			
77	Ir		7			2			
78	Pt		9			1			
79	Au		10			1			
80	Hg	$1s$ to $5d$ Filled				2			
81	Tl					2	1		
82	Pb					2	2		
83	Bi					2	3		
84	Po					2	4		
85	At					2	5		
86	Rn			0	0	2	6		
87	Fr					$6s$ and $6p$ Filled			1
88	Ra								2
89	Ac							1	2
90	Th							2	2
91	Pa			2				1	2
92	U			3				1	2
93	Np			5				0	2
94	Pu			6				0	2
95	Am			7				0	2
96	Cm			7				1	2
97	Bk			8				1	2
98	Cf			10				0	2
99	Es			11				0	2
100	Fm			12				0	2
101	Md			13				0	2
102	No			14				0	2
103	Lw			14				1	2

Appendix II Selected Properties of the Elements

Atomic weights given as 1961 atomic weights based on carbon-12 rounded off to a maximum of three decimal places.

Atomic weights in parentheses are those of the most stable isotopes.

ELEMENT	SYMBOL	ATOMIC NUMBER	ATOMIC WEIGHT	MELTING POINT, °C	DENSITY OF SOLID G/CC	CRYSTAL STRUCTURE OF SOLID
Actinium	Ac	89	(227)	1050		FCC
Aluminum	Al	13	26.982	660	2.70	FCC
Americium	Am	95	(243)		11.7	Hexag
Antimony	Sb	51	121.75	630.5	6.68	Rhomb
Argon	Ar	18	39.948	−189.4	1.67	FCC
Arsenic	As	33	74.922	817[a]	5.72	Rhomb*
Astatine	At	85	(210)	~300		
Barium	Ba	56	137.34	714	3.5	BCC*
Berkelium	Bk	97	(247)			
Beryllium	Be	4	9.012	1277	1.85	HCP*
Bismuth	Bi	83	208.980	271	9.80	Rhomb
Boron	B	5	10.811	2030	2.34	Tetrag*
Bromine	Br	35	79.909	−7.2		Orthorho
Cadmium	Cd	48	112.40	321	8.65	HCP*
Calcium	Ca	20	40.08	850	1.55	FCC*
Californium	Cf	98	(249)			
Carbon	C	6	12.011	3727[b]	3.51[c]	DC*
Cerium	Ce	58	140.12	804	6.77	FCC*
Cesium	Cs	55	132.905	28.7	1.90	BCC
Chlorine	Cl	17	35.453	−101	1.9	Tetrag
Chromium	Cr	24	51.996	1875	7.19	BCC*
Cobalt	Co	27	58.933	1495	8.85	HCP*
Copper	Cu	29	63.54	1083	8.96	FCC
Curium	Cm	96	(247)			
Dysprosium	Dy	66	162.50	1407	8.55	HCP
Einsteinium	Es	99	(254)			
Erbium	Er	68	167.26	1497	9.15	HCP
Europium	Eu	63	151.96	826	5.24	BCC

Appendix II Con't

ELEMENT	SYMBOL	ATOMIC NUMBER	ATOMIC WEIGHT	MELTING POINT, °C	DENSITY OF SOLID G/CC	CRYSTAL STRUCTURE OF SOLID
Fermium	Fm	100	(253)			
Fluorine	F	9	18.998	−220	1.3	
Francium	Fr	87	(223)	~27		BCC
Gadolinium	Gd	64	157.25	1312	7.86	HCP*
Gallium	Ga	31	69.72	29.8	5.91	Orthorho
Germanium	Ge	32	72.59	937	5.32	DC
Gold	Au	79	196.967	1063	19.32	FCC
Hafnium	Hf	72	178.49	2222	13.1	HCP*
Helium	He	2	4.003	−269.7		HCP
Holmium	Ho	67	164.930	1461	6.79	HCP
Hydrogen	H	1	1.008	−259.2		Hexag
Indium	In	49	114.82	156.2	7.31	FC Tetrag
Iodine	I	53	126.904	113.7	4.94	Orthorho
Iridium	Ir	77	192.2	2454	22.4	FCC
Iron	Fe	26	55.847	1537	7.87	BCC*
Krypton	Kr	36	83.80	−157.3	3.0	FCC
Lanthanum	La	57	138.91	920	6.19	Hexag*
Lawrencium	Lw	103	(257)			
Lead	Pb	82	207.19	327.3	11.34	FCC
Lithium	Li	3	6.939	180.5	0.53	BCC
Lutetium	Lu	71	174.97	1650	9.85	HCP
Magnesium	Mg	12	24.312	650	1.74	HCP
Manganese	Mn	25	54.938	1245	7.43	Cubic*
Mendelevium	Mv	101	(256)			
Mercury	Hg	80	200.59	−38.4	14.19	Rhomb
Molybdenum	Mo	42	95.94	2610	10.2	BCC
Neodymium	Nd	60	144.24	1019	7.0	Hexag*
Neon	Ne	10	20.183	−248.7	1.45	FCC
Neptunium	Np	93	(237)	637	19.5	Orthorho*
Nickel	Ni	28	58.71	1453	8.9	FCC
Niobium	Nb	41	92.906	2415	8.6	BCC
Nitrogen	N	7	14.007	−210	1.03	Hexag*
Nobelium	No	102	(253)			
Osmium	Os	76	190.2	2700	22.57	HCP
Oxygen	O	8	15.999	−218.8	1.43	Cubic*
Palladium	Pd	46	106.4	1552	12.02	FCC
Phosphorus[d]	P	15	30.974	44.2	1.83	Cubic*
Platinum	Pt	78	195.09	1769	21.45	FCC
Plutonium	Pu	94	(242)	640	19.3	Monoclinic*

ELEMENT	SYMBOL	ATOMIC NUMBER	ATOMIC WEIGHT	MELTING POINT, °C	DENSITY OF SOLID G/CC	CRYSTAL STRUCTURE OF SOLID
Polonium	Po	84	(210)	254	9.2	Monoclinic*
Potassium	K	19	39.102	63.7	0.86	BCC
Praseodymium	Pr	59	140.907	919	6.77	Hexag*
Promethium	Pm	61	(147)	1027		Hexag
Protoactinium	Pa	91	(231)	1230	15.4	BC Tetrag
Radium	Ra	88	(226)	700	5.0	BC Tetrag
Radon	Rn	86	(222)	−71	4	FCC
Rhenium	Re	75	186.2	3180	21.04	HCP
Rhodium	Rh	45	102.905	1966	12.44	FCC*
Rubidium	Rb	37	85.47	38.9	1.53	BCC
Ruthenium	Ru	44	101.07	2500	12.2	HCP*
Samarium	Sm	62	150.35	1072	7.49	Rhomb*
Scandium	Sc	21	44.956	1539	2.99	HCP*
Selenium	Se	34	78.96	217	4.79	Hexag*
Silicon	Si	14	28.086	1410	2.34	DC
Silver	Ag	47	107.870	960.8	10.5	FCC
Sodium	Na	11	22.990	97.8	0.97	BCC
Strontium	Sr	38	87.62	768	2.60	FCC*
Sulfur[e]	S	16	32.064	119	2.07	Orthorho*
Tantalum	Ta	73	180.948	2996	16.6	BCC
Technetium	Tc	43	(99)	2200	11.5	HCP
Tellurium	Te	52	127.60	450	6.24	Hexag
Terbium	Tb	65	158.924	1356	8.25	HCP
Thallium	Tl	81	204.37	303	11.85	HCP*
Thorium	Th	90	232.038	1750	11.66	FCC*
Thulium	Tm	69	168.934	1545	9.31	HCP
Tin	Sn	50	118.69	231.9	7.30	Tetrag*
Titanium	Ti	22	47.90	1668	4.51	HCP*
Tungsten	W	74	183.85	3410	19.3	BCC
Uranium	U	92	238.03	1132	19.05	Orthorho*
Vanadium	V	23	50.942	1900	6.1	BCC
Xenon	Xe	54	131.30	−112	3.6	FCC
Ytterbium	Yb	70	173.04	824	6.96	FCC*
Yttrium	Y	39	88.905	1509	4.47	HCP*
Zinc	Zn	30	65.37	419.5	7.13	HCP
Zirconium	Zr	40	91.22	1852	6.49	HCP*

[a] Melting point at 28 atm pressure
[b] Sublimes
[c] Diamond
[d] White phosphorus
[e] Yellow sulfur
* Other crystal modifications exist

Appendix III Crystallographic Indices

Although the description of atom positions in a unit cell is a complete description of the crystal structure, it is useful to have a means of describing the planes and directions in a crystal as well. For this purpose, a system of crystallographic indices has been developed.

INDICES OF A LATTICE PLANE

A plane in space satisfies the equation

$$\frac{x}{a} + \frac{y}{b} + \frac{z}{c} = 1$$

where a, b, and c are the intercepts of the plane on the x, y, and z axes respectively. In describing crystallographic planes, the axes are taken along three nonparallel edges of the unit cell, and the intercepts are measured in terms of a *unit length,* which is assigned to each edge of the cell regardless of its actual dimension. The use of the intercepts a, b, and c to represent a crystallographic plane inside a unit cell has the disadvantage that the intercepts are often fractions less than one and may be infinite (for a plane parallel to an axis). For this reason it is now common practice to use the reciprocals of the intercepts to designate a crystallographic plane.

$$h = \frac{1}{a}, \quad k = \frac{1}{b}, \quad l = \frac{1}{c} \qquad \text{so that } hx + ky + lz = 1 \qquad \text{(III.1)}$$

The *indices of a plane* (usually called *Miller indices*) are simply h, k, and l enclosed in parentheses, (hkl), and are used to describe the plane. For example, in a cubic unit cell, that face of the cube

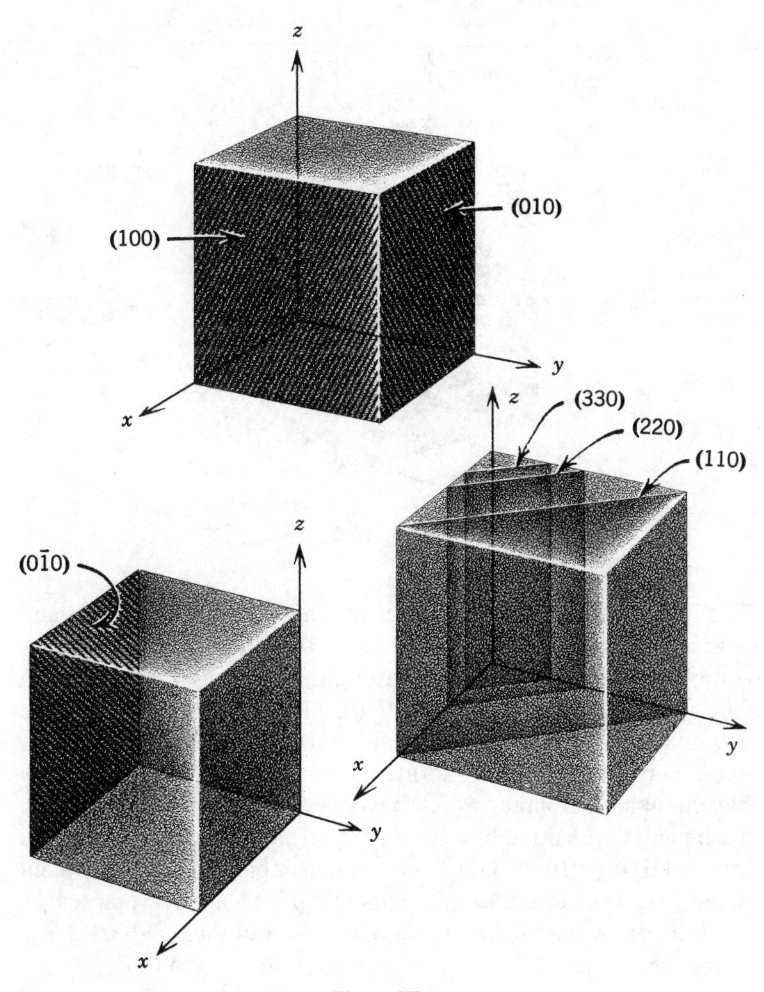

Figure III.1

which intersects the positive x axis is the (100) plane, as shown in Figure III.1. The indices of the opposite face of the cube can be determined by translating the origin of the coordinate system to another corner of the unit cell. (This is permissible, for all corners of the unit cell are lattice points and are therefore equivalent.) The plane then intersects the negative x axis; to indicate this, a minus sign is placed above the reciprocal of the intercept: ($\bar{1}$00).

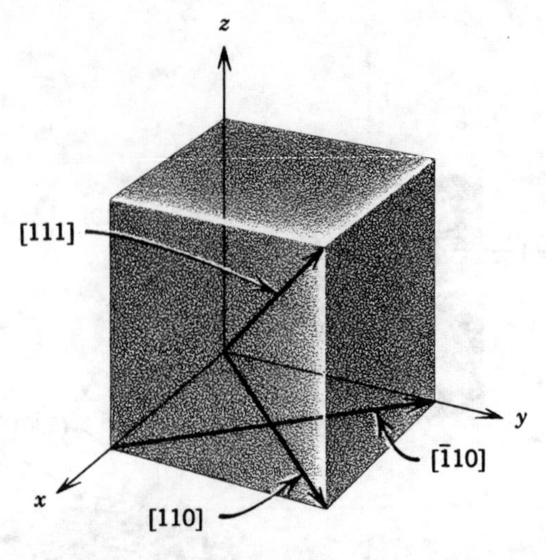

Figure III.2

The indices of parallel planes on the same side of the origin and with identical atomic distributions are often reduced to the smallest common indices by dividing through with the largest common factor; thus the ($\frac{1}{3}$ $\frac{1}{3}$ 0), and ($\frac{1}{2}$ $\frac{1}{2}$ 0), planes would all be called 110 planes unless there is some particular reason to distinguish between them. In addition, crystals with high symmetry have nonparallel planes which have identical atomic distributions. Such planes are said to be *crystallographically equivalent*. Examples are the (100), (010), and (001) planes in a cubic lattice. Equivalent planes are said to be of the same *form* and are represented by the indices of one of the planes, enclosed in braces: {100} designates the form, (100), (010), (001), ($\bar{1}$00), (0$\bar{1}$0), and (00$\bar{1}$), the six cube faces.

INDICES OF A LATTICE DIRECTION

The *indices of a direction* are simply the vector components of the direction resolved along each of the coordinate axes and reduced to smallest integers. For example, in a cubic unit cell, if the origin is at the corner and the axes are parallel to the edges, the body diagonal would be represented as [111], Figure III.2.

The numbers "111" are the indices of the direction, and the use of brackets rather than parentheses denotes that they refer to a crystallographic direction rather than to a plane. One of the face diagonals would be a [110] direction. Directions of the same linear atomic distribution are said to be of the same *form* and have indices set off by carats: $\langle 110 \rangle$ includes [110], $[10\bar{1}]$, $[0\bar{1}\bar{1}]$, etc. It is useful to note that in the cubic system, and *only in the cubic system,* all directions and planes with identical indices are perpendicular. That is, the normal to a plane is a direction with the same indices as the plane. This relationship permits the use of vector algebra to simplify many crystallographic calculations.

VECTOR ALGEBRA FOR CUBIC LATTICES

Since the crystallographic directions and normals to planes in a cubic lattice are both vectors, the techniques of vector multiplication may be applied to them. Vectors may be multiplied to produce either a *dot* (*scalar*) *product* or a *cross* (*vector*) *product.*

The dot product of the vectors **A** and **B** is written $\mathbf{A} \cdot \mathbf{B}$ and is a scalar which is the product of the magnitude of one vector (**A**) and the magnitude of the component of the second vector (**B**) resolved in the direction of **A**. Thus,

$$\mathbf{A} \cdot \mathbf{B} = AB \cos \varphi \tag{III.2}$$

where φ is the angle between the vectors. If **A** and **B** are resolved into components along the x, y, and z axes,

$$\begin{aligned} \mathbf{A} &= A_1\mathbf{u} + A_2\mathbf{v} + A_3\mathbf{w} \\ \mathbf{B} &= B_1\mathbf{u} + B_2\mathbf{v} + B_3\mathbf{w} \end{aligned} \tag{III.3}$$

where **u**, **v**, and **w** are unit vectors in the x, y, and z directions respectively. Then

$$\begin{aligned} \mathbf{A} \cdot \mathbf{B} &= (A_1\mathbf{u} + A_2\mathbf{v} + A_3\mathbf{w}) \cdot (B_1\mathbf{u} + B_2\mathbf{v} + B_3\mathbf{w}) \\ &= A_1B_1 + A_2B_2 + A_3B_3 \end{aligned} \tag{III.4}$$

since $\mathbf{u} \cdot \mathbf{u} = 1$, etc. In terms of the indices of two directions (which can be treated as vectors with arbitrary magnitude),

$$[hkl] \cdot [h'k'l'] = hh' + kk' + ll' \tag{III.5}$$

and, in particular, *the dot product of two perpendicular directions is zero.*

The cross product of **A** and **B**, written $\mathbf{A} \times \mathbf{B}$, is a vector with a magnitude given by

$$\mathbf{A} \times \mathbf{B} = \mathbf{C} = AB \sin \varphi \tag{III.6}$$

where φ is the angle between **A** and **B**. The direction of **C** is normal to the plane of **A** and **B** and is positive if rotating **A** into **B** would produce motion along **C** in a right-hand screw. Therefore cross-product multiplication is not commutative.

$$\begin{aligned} \mathbf{A} \times \mathbf{B} &= -(\mathbf{B} \times \mathbf{A}) \\ \mathbf{u} \times \mathbf{v} &= -(\mathbf{v} \times \mathbf{u}) = \mathbf{w} \\ \mathbf{v} \times \mathbf{w} &= -(\mathbf{w} \times \mathbf{v}) = \mathbf{u} \\ \mathbf{w} \times \mathbf{u} &= -(\mathbf{u} \times \mathbf{w}) = \mathbf{v} \end{aligned} \tag{III.7}$$

Also $\mathbf{u} \times \mathbf{u} = 0$, etc.

If **A** and **B** are resolved into components and a cross product is taken,

$$\begin{aligned} \mathbf{A} \times \mathbf{B} &= (A_1\mathbf{u} + A_2\mathbf{v} + A_3\mathbf{w}) \times (B_1\mathbf{u} + B_2\mathbf{v} + B_3\mathbf{w}) \\ &= A_1B_2\mathbf{w} - A_1B_3\mathbf{v} - A_2B_1\mathbf{w} + A_2B_3\mathbf{u} + A_3B_1\mathbf{v} - A_3B_2\mathbf{u} \\ &= (A_2B_3 - A_3B_2)\mathbf{u} + (\mathrm{A}_3B_1 - A_1B_3)\mathbf{v} + (A_1B_2 - A_2B_1)\mathbf{w} \end{aligned} \tag{III.8}$$

which is the same as the expansion of a third-order determinant. Therefore the cross product can also be written

$$\mathbf{A} \times \mathbf{B} = \begin{vmatrix} \mathbf{u} & \mathbf{v} & \mathbf{w} \\ A_1 & A_2 & A_3 \\ B_1 & B_2 & B_3 \end{vmatrix} \tag{III.9}$$

In terms of direction indices,

$$[hkl] \times [h'k'l'] = \begin{vmatrix} \mathbf{u} & \mathbf{v} & \mathbf{w} \\ h & k & l \\ h' & k' & l' \end{vmatrix} \tag{III.10}$$

This relationship is commonly used to find the direction of the line of intersection of two planes in a cubic crystal: *the line of intersection is perpendicular to the normals of both planes, and its direction is therefore the cross product of the two normals.*

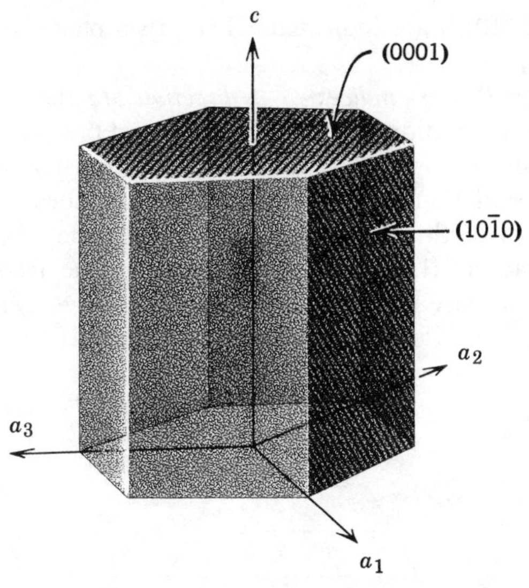

Figure III.3

MILLER-BRAVAIS INDICES

An alternate indexing system, which has four numbers in each set of indices, is often used for hexagonal crystals. The indices are called *Miller-Bravais indices.* Four numbers are used in order to make the relationship between the indices and the symmetry of the hexagonal lattice more obvious. If the C cell in Figure 3.1 is used for the hexagonal unit cell, it is described with reference to four axes, one along the axis of the hexagonal prism and three in the base, 120° apart (see Figure III.3).

The *Miller-Bravais indices of a plane* are denoted by h, k, i, and l enclosed in parentheses, $(hkil)$. These indices are the reciprocals of intercepts on the a_1, a_2, a_3, and c axes, respectively. As with Miller indices, the reciprocals are usually divided by the largest common factor. Since only three noncoplanar axes are necessary to specify a plane in space, the four indices cannot be independent. The additional condition which their values must satisfy is

$$h + k = -i \qquad \text{(III.11)}$$

In Figure III.3, the (0001) plane—called the basal plane—and the prism plane $(10\bar{1}0)$ are indicated. The prism plane is one of the form $\{10\bar{1}0\}$.

The *Miller-Bravais indices of a direction* are the vector components of the direction, resolved along each of the four coordinate axes and reduced to smallest integers. For consistency, the same restriction (Equation III.11) which applies to planes is applied to directions. The three coplanar components may be made to satisfy Equation III.11 by drawing them on the triangular net shown on the base of the prism in Figure III.4*a*. The $[10\bar{1}0]$, $[11\bar{2}0]$ and $[10\bar{1}1]$ directions are drawn in Figure III.4*b*.

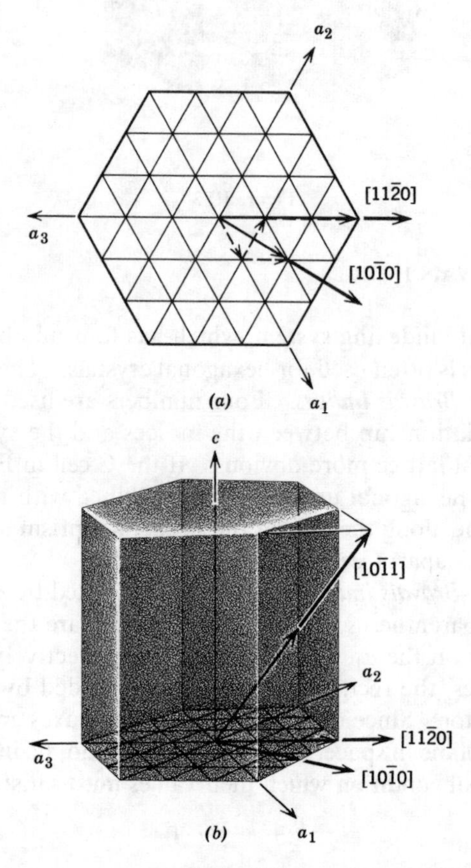

Figure III.4

Appendix IV The Diffraction of X-rays by Cubic Crystals

The analysis of X-ray diffraction patterns is a complicated process for crystals with low symmetry. However, the principles and rationale of the analysis can be described relatively simply for the case of crystals having high symmetry, namely cubic crystals.

When X-rays impinge on a crystal, they are scattered by the atoms. Since the scattered radiation results from the acceleration and deceleration of electrons set in motion by the X-rays, it has the same wavelength as the incident X-rays. This fact, plus the regularity of the pattern of atoms in a crystal, permits the treatment of a crystal as a three-dimensional diffraction grating. Diffraction, or constructive interference, occurs only when the difference in the distance traveled by two identical diffracted waves is an integral number of wavelengths, so that the two waves are in phase. In Figure IV.1 the horizontal lines represent atomic planes of interplanar spacing d, the atoms of which are the scattering centers for the incident radiation. The total path difference between the two rays shown is $2d \sin \theta$; therefore the equation

$$n\lambda = 2d \sin \theta \tag{IV.1}$$

describes the conditions under which diffraction occurs. Equation IV.1 is known as Bragg's Law of X-ray diffraction; in it, n may be any integer. Physically, the equation states that λ, d, and θ must have values which will yield integral values of n in order for an intensity peak to be observed.

In the *powder method* of X-ray diffraction, a sample is ground to a powder in order to expose all possible orientations of the crystal to the X-ray beam. Proper values of λ, d, and θ for diffraction are achieved as follows:

1. λ is kept constant by using filtered X-radiation that is approximately *monochromatic.*
2. d may have any value consistent with the crystal structure.

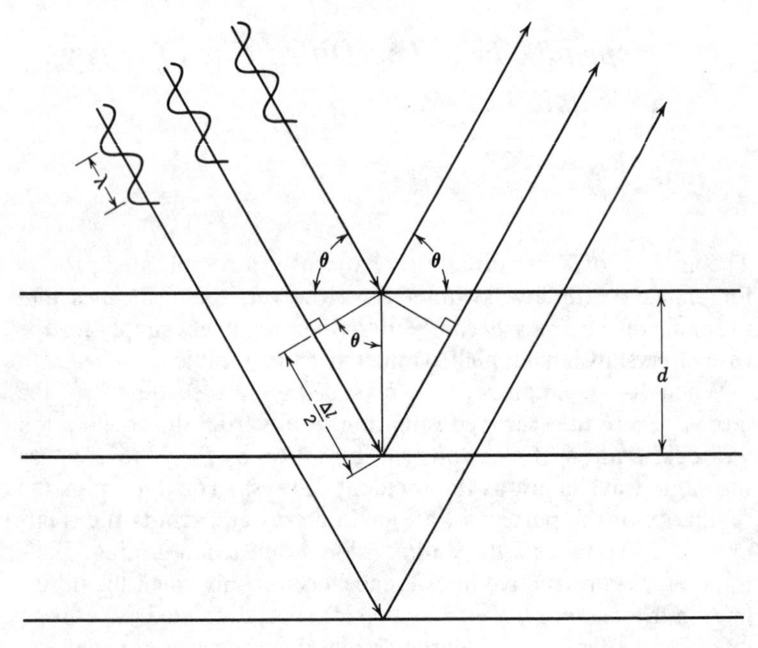

Figure IV.1

3. θ is the variable parameter, in terms of which the diffraction peaks are measured.

Because of destructive interference, intensity peaks cannot occur at angles which do not satisfy the Bragg Law. However, they do not necessarily occur when the Bragg Law is satisfied. The reason for the latter statement is as follows. The intensity of a diffracted X-ray beam depends primarily on two properties of the crystal, the *atomic scattering factor* of each atom and the *position of each atom in the unit cell.* The atomic scattering factor f is a measure of the efficiency of cooperation of all the electrons surrounding a nucleus in scattering the incident radiation. It is a function of θ; if θ is small, f approaches the atomic number of the element, for the electrons scatter rays which are nearly in phase; as θ increases, f decreases because the waves from individual electrons are more out of phase. In practice, values of f are tabulated as functions of $\sin \theta/\lambda$ for each of the elements. The higher the atomic number of an element, the more effective a scatterer it will be. The second important factor is the intensity arising from the cooperation of the various atoms in the unit cell in scattering the incident radiation. *Each atom in the cell scatters with an amplitude proportional to its value of f, and with a phase which is dependent on its position in the cell.* For the same value of λ, differences in scattering power of different atoms produce various amplitudes and phases in the radiation they scatter, and thus to obtain a measure of the cooperative effort we must add sine waves of different amplitudes and phases. This addition is accomplished in the following way. Let the amplitude of each wave be represented by the length of a vector, and let its phase be represented by the direction ϕ of the vector. Vectors $\mathbf{f}_1$, $\mathbf{f}_2$, $\mathbf{f}_3$, and $\mathbf{f}_4$ in Figure IV.2*a* represent atomic

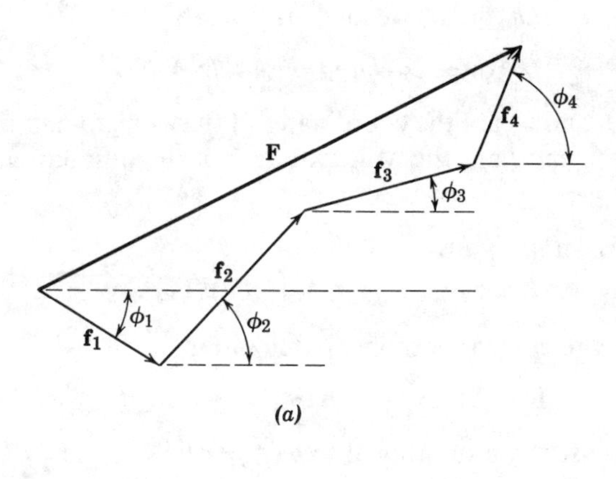

(a)

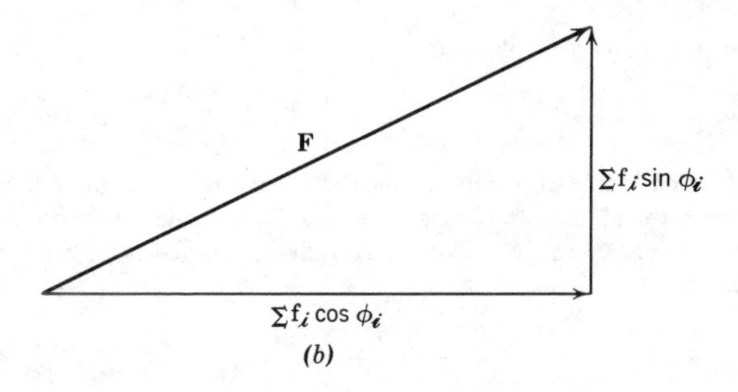

(b)

Figure IV.2

scattering factors for four atoms in a unit cell. **F** is the resultant vector, and $|\mathbf{F}|^2$ will be proportional to the intensity of scattered radiation observed in a direction given by the direction of **F**. Each individual vector can be resolved into horizontal and vertical components, as illustrated in Figure IV.2*b*. Then

$$|\mathbf{F}|^2 = (\Sigma f_i \cos \phi_i)^2 + (\Sigma f_i \sin \phi_i)^2 \qquad \text{(IV.2)}$$

More commonly, the problem is treated analytically, rather than geometrically, by using complex number notation.

$$\mathbf{F} = \Sigma f_i(\cos \phi_i + i \sin \phi_i) \qquad \text{(IV.3)}$$

$$\mathbf{F}^* = \Sigma f_i(\cos \phi_i - i \sin \phi_i) \qquad \text{(IV.4)}$$

$$\mathbf{FF}^* = |\mathbf{F}|^2 = (\Sigma f_i \cos \phi_i)^2 + (\Sigma f_i \sin \phi_i)^2 \qquad \text{(IV.5)}$$

where $\mathbf{F}^*$ is the complex conjugate of **F**.

In a *cubic crystal,* the phase angle is given by[1]

$$\phi_i = 2\pi(hu_i + kv_i + lw_i) \qquad \text{(IV.6)}$$

where u_i, v_i, and w_i are the coordinates of the i-th atom in the unit cell and h, k, and l are the Miller indices of the diffracting plane. Therefore

$$\mathbf{F} = \Sigma f_i[\cos 2\pi(hu_i + kv_i + lw_i) + i \sin 2\pi(hu_i + kv_i + lw_i)] \qquad \text{(IV.7)}$$

or, in the more compact exponential notation,

$$\mathbf{F} = \Sigma f_i \exp[2\pi i(hu_i + kv_i + lw_i)] \qquad \text{(IV.8)}$$

and the intensity is proportional to $\mathbf{FF}^* = |\mathbf{F}|^2$.

[1] The general equation for a phase angle is

$$\phi_i = 2\pi\frac{\delta}{\lambda},$$

where δ is the path difference between the diffracted beam from the i-th atom and the beam from the atom at the origin. The one-dimensional case, where d is parallel to x, is established easily, since the path difference is just twice the interatomic distance and $\sin^2 \phi = 1$.

$$\phi_i = 2\pi\frac{\delta}{\lambda} = \frac{2\pi u_i a_0}{d}$$

By the definition of Miller indices, $d = a_0/h$, and so $\phi_i = 2\pi(hu_i)$. Equation IV.6 is derived by considering the problem in three dimensions.

SIMPLE CUBIC CRYSTALS

A simple cubic crystal has only one atom per unit cell; therefore

$$|\mathbf{F}|^2 = f_1^2 \tag{IV.9}$$

and all the lines predicted by the Bragg Law will occur if enough random crystal orientations are present and if the atomic scattering factor is large enough that the intensity peaks can be seen above the background radiation.

BODY-CENTERED CUBIC CRYSTALS

A body-centered cubic crystal has two atoms per unit cell: atom 1 with coordinates 0, 0, 0 and atom 2 with coordinates $\frac{1}{2}, \frac{1}{2}, \frac{1}{2}$.

$$\mathbf{F} = \Sigma f_i \exp\left[2\pi i(hu_i + kv_i + lw_i)\right] \tag{IV.10}$$

Therefore

$$|\mathbf{F}|^2 = [f_1 + f_2 \cos \pi i\,(h + k + l)]^2 \tag{IV.11}$$

because $\sin(n\pi) = 0$ when n is any integer.

Two cases must be distinguished:

1. If $(h + k + l)$ is even, intensity is proportional to $(f_1 + f_2)^2$.
2. If $(h + k + l)$ is odd, intensity is proportional to $(f_1 - f_2)^2$. Thus, for $(h + k + l)$ odd, lines will be of comparatively low intensity, and as the values of f_1 and f_2 approach each other, the intensity approaches zero. In particular, when *both atoms are identical,* f_1 equals f_2 and
3. If $(h + k + l)$ is even, intensity is proportional to $4f^2$.
4. If $(h + k + l)$ is odd, intensity is zero.

FACE-CENTERED CUBIC CRYSTALS

A face-centered cubic crystal has four atoms per unit cell:

Atom	*Coordinates*
1	0, 0, 0
2	$\frac{1}{2}$, 0, $\frac{1}{2}$
3	$\frac{1}{2}$, $\frac{1}{2}$, 0
4	0, $\frac{1}{2}$, $\frac{1}{2}$

Therefore

$$|\mathbf{F}|^2 = [f_1 + f_2 \cos \pi(h + l) + f_3 \cos \pi(h + k) + f_4 \cos \pi(k + l)]^2 \qquad \text{(IV.12)}$$

Let us consider two different situations:

1. All atoms are alike; $f_1 = f_2 = f_3 = f_4$.
   a. If h, k, l are all odd or all even, intensity is proportional to $16f^2$.
   b. If h, k, l are mixed, intensity equals zero.
2. Two atoms of each kind, A and B.
   a. If h, k, l are all odd or all even, irrespective of which ones occupy the four possible positions, intensity equals $4(f_A + f_B)^2$.
   b. If h, k, l are mixed, the table below gives the intensities for the possible combinations.

POSITIONS OF A AND B ATOMS	h, k ODD; l EVEN h, k EVEN; l ODD	h, l ODD; k EVEN h, l EVEN; k ODD	k, l ODD; h EVEN k, l EVEN; h ODD
A(1 & 2), B(3 & 4)	0	$4(f_A - f_B)^2$	0
A(1 & 3), B(2 & 4)	$4(f_A - f_B)^2$	0	0
A(1 & 4), B(2 & 3)	0	0	$4(f_A - f_B)^2$

DETERMINING CUBIC CRYSTAL STRUCTURES

The analytical procedure for determining a crystal structure is to index the observed diffraction peaks and then to determine the crystal structure from a knowledge of which planes do and do not produce peaks. The following relations simplify the procedure for some of the less complicated cubic crystal structures.

CRYSTAL STRUCTURE	INDICES OF REFLECTING PLANES
Simple cubic	all values of $(h^2 + k^2 + l^2)$ except 7, 15, 23 . . .
Body-centered cubic	$h + k + l$ is even (no peak when $h + k + l$ is odd)
Face-centered cubic	h, k, l all even or all odd (unmixed) (no peak when h, k, l are mixed)
Diamond cubic	$h + k + l$ is odd or an even multiple of 2 (no peak when $h + k + l$ is an odd multiple of 2)

The observed peaks are indexed by writing *d*, the interplanar spacing in the Bragg equation, in terms of the Miller indices of the reflecting planes and the lattice parameter a_0.

$$d^2 = \frac{{a_0}^2}{(h^2 + k^2 + l^2)} \tag{IV.13}$$

Combining the above equation with the Bragg equation,

$$\sin^2 \theta = \frac{\lambda^2}{4{a_0}^2}(h^2 + k^2 + l^2) \tag{IV.14}$$

where n^2 has been absorbed into the indices of the planes (a second-order (110) reflection cannot be distinguished from a first-order (220) reflection). Since λ and a_0 are constant, θ depends only on h, k, and l and can be indexed accordingly.

$$\sin^2 \theta = C(h^2 + k^2 + l^2) \tag{IV.15}$$

where C is a constant, $\lambda^2/4{a_0}^2$. Integers are then found for h, k, and l that satisfy the above equation for the angles of *all* the observed diffraction peaks. The lattice parameter is then calculated from the value of C and the wavelength λ.

Appendix V Structural Formulae of Selected Polymers

Polymer	Formula
Polyethylene	see below
Polyvinyl chloride	see below
Polyvinylidene chloride	see below
Polypropylene	see below

Polyethylene

```
  H   H   H   H   H   H   H   H   H   H   H   H
  |   |   |   |   |   |   |   |   |   |   |   |
 -C---C---C---C---C---C---C---C---C---C---C---C-
  |   |   |   |   |   |   |   |   |   |   |   |
  H   H   H   H   H   H   H   H   H   H   H   H
```

Polyvinyl chloride

```
  H   H   H   H   H   Cl  H   H   H   Cl  H   Cl
  |   |   |   |   |   |   |   |   |   |   |   |
 -C---C---C---C---C---C---C---C---C---C---C---C-
  |   |   |   |   |   |   |   |   |   |   |   |
  H   Cl  H   Cl  H   H   H   Cl  H   H   H   H
```

Polyvinylidene chloride

```
  H   Cl  H   Cl  H   Cl  H   Cl  H   Cl  H   Cl
  |   |   |   |   |   |   |   |   |   |   |   |
 -C---C---C---C---C---C---C---C---C---C---C---C-
  |   |   |   |   |   |   |   |   |   |   |   |
  H   Cl  H   Cl  H   Cl  H   Cl  H   Cl  H   Cl
```

Polypropylene

```
                  H         H         H                   H
                  |         |         |                   |
  H   H   H   H---C---H H H-C-H   H H-C-H   H   H   H   H-C-H
  |   |   |       |     | | |     | | |     |   |   |     |
 -C---C---C-------C-----C-C-------C-C-------C---C---C-----C-
  |   |   |       |     | |       | |       |   |   |     |
  H H-C-H H       H     H H       H H       H H-C-H H     H
      |                                         |
      H                                         H
```

Polystyrene

Polyvinyl acetate

Polymer	Formula
Polyvinyl alcohol	$—CH_2—CH(OH)—CH_2—CH(OH)—CH_2—CH(OH)—CH_2—CH(OH)—CH_2—CH(OH)—CH_2—CH(OH)—$
Polymethyl methacrylate	$—CH_2—C(CH_3)(COOCH_3)—CH_2—C(CH_3)(COOCH_3)—CH_2—C(COOCH_3)(CH_3)—CH_2—C(CH_3)(COOCH_3)—CH_2—C(COOCH_3)(CH_3)—CH_2—C(CH_3)(COOCH_3)—$
Polychloroprene (Neoprene)	$—CH_2—C(Cl)=CH—CH_2—CH_2—C(Cl)(CH=CH_2)—CH_2—C(Cl)=CH—CH_2—CH_2—C(Cl)(CH=CH_2)—$

Polymer	Structural formula
cis-Polyisoprene (natural rubber)	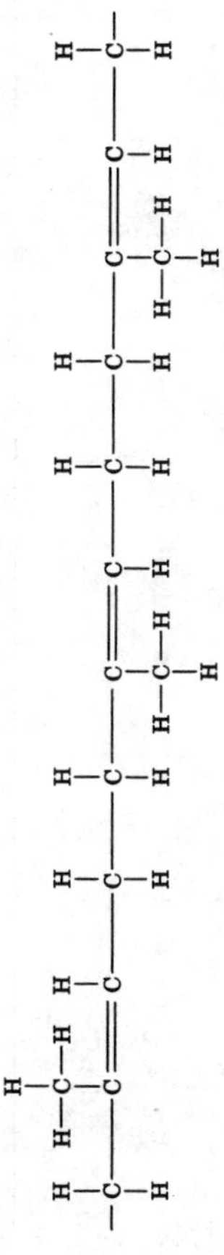 $-CH_2-C(CH_3)=CH-CH_2-CH_2-C(CH_3)=CH-CH_2-CH_2-C(CH_3)=CH-CH_2-$
trans-Polyisoprene (gutta percha)	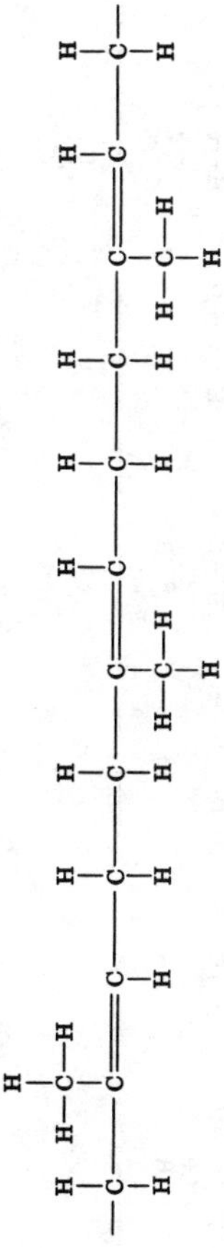 $-CH_2-C(CH_3)=CH-CH_2-CH_2-C(CH_3)=CH-CH_2-CH_2-C(CH_3)=CH-CH_2-$
Polybutadiene	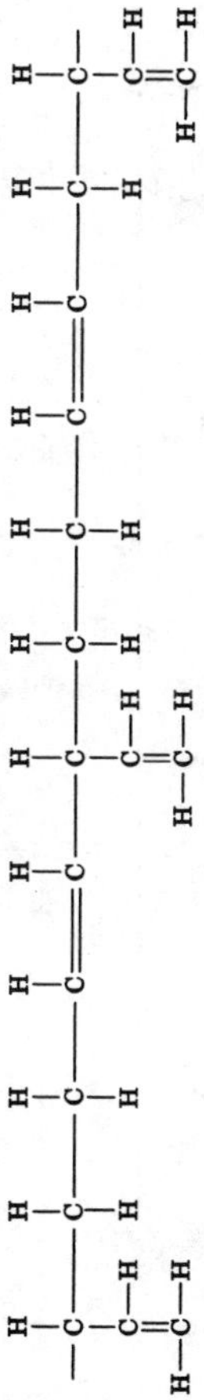 $-CH(CH=CH_2)-CH_2-CH_2-CH=CH-CH(CH=CH_2)-CH_2-CH_2-CH=CH-CH_2-CH(CH=CH_2)-$
Polyhexamethylene adipamide (Nylon 6, 6)	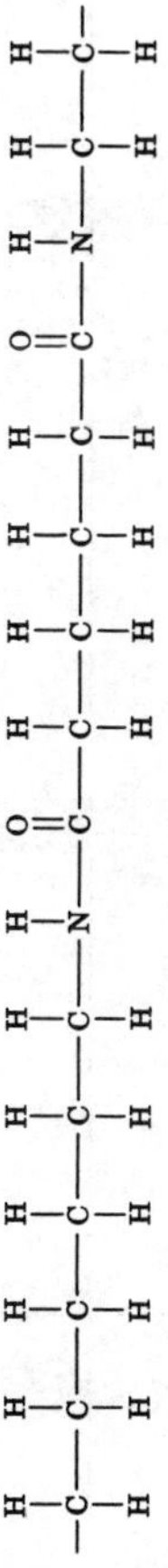 $-CH_2-CH_2-CH_2-CH_2-CH_2-CH_2-NH-C(=O)-CH_2-CH_2-CH_2-CH_2-C(=O)-NH-CH_2-CH_2-$

Polymer	Formula
Polytetrafluoroethylene (Teflon)	F F F F F F F F F F F F —C—C—C—C—C—C—C—C—C—C—C—C— F F F F F F F F F F F F
Methyl silicone	H H H H H H H—C—H H—C—H H—C—H H—C—H H—C—H H—C—H —Si—O—Si—O—Si—O—Si—O—Si—O—Si— H—C—H H—C—H H—C—H H—C—H H—C—H H—C—H H H H H H H
Phenol–formaldehyde (Bakelite)	H—C—H H—C—H OH OH H H H—C C—H H OH H H—C C—H H —C—C C—C—C C—C—C C—C—C C—C— H H—C C—H H C H H—C C—H H C H H—C—H OH H—C—H OH

Polymer	Structure
Urea formaldehyde	H H H H H H —N—C—N—C—N—C—N—C—N—C—N—C— C=O H H C=O H H C=O H H
Cellulose	H OH CH_2OH H OH O C C C O O C C OH H H H H OH H C C C C C C H H OH H H H H H C O O C C C O CH_2OH H OH CH_2OH
Cellulose nitrate (Nitrocellulose)	H NO_3 CH_2NO_3 H NO_3 O C C C O O C C OH H H H H OH H C C C C C C H H OH H H H H H C O O C C C O CH_2NO_3 H NO_3 CH_2NO_3

Appendix VI Selected Equilibrium Diagrams

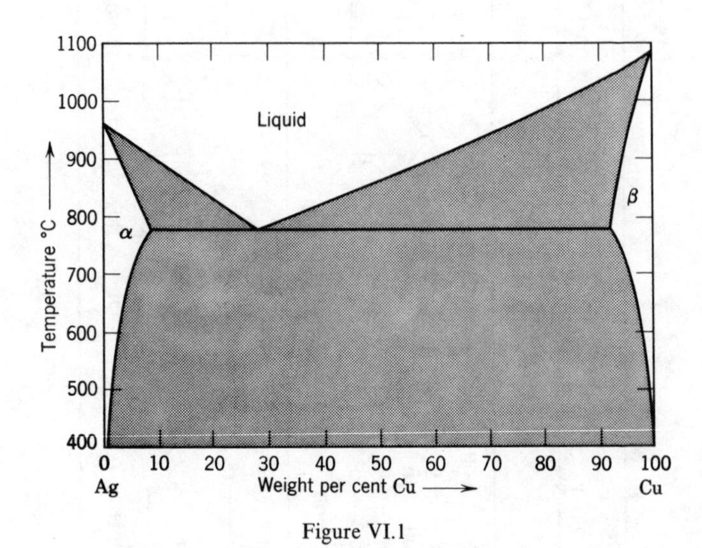

Figure VI.1

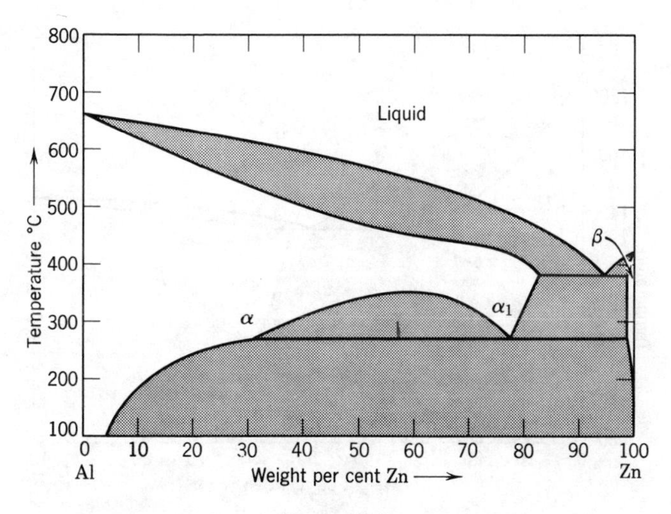

Figure VI.2

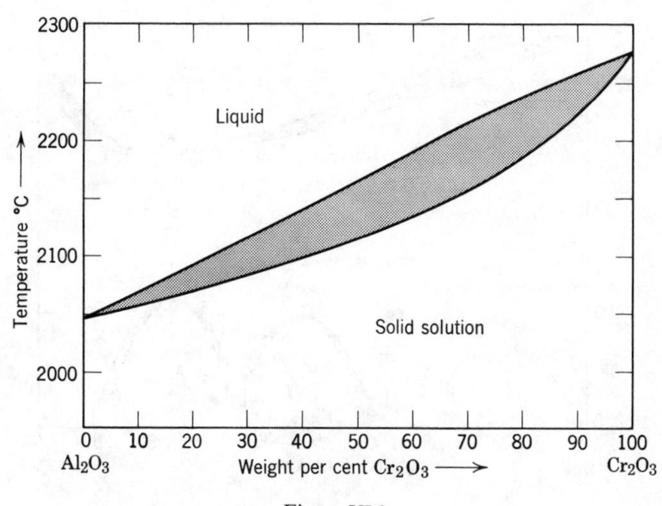

Figure VI.3

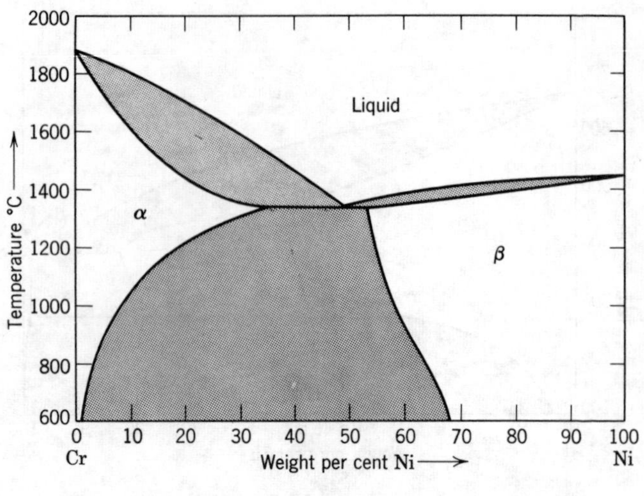

Figure VI.4

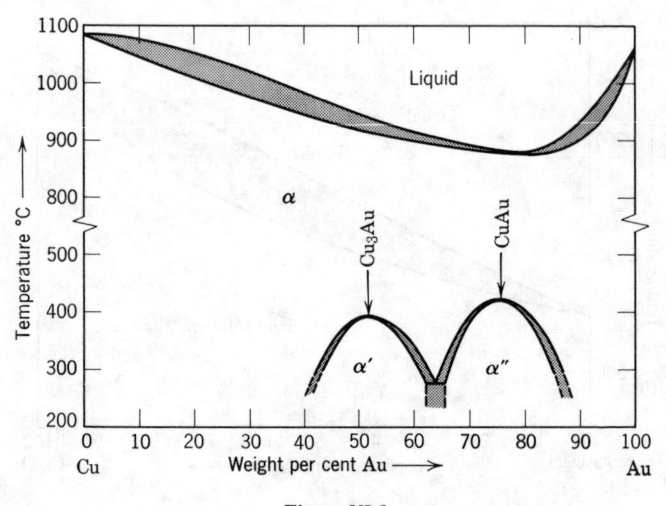

Figure VI.5

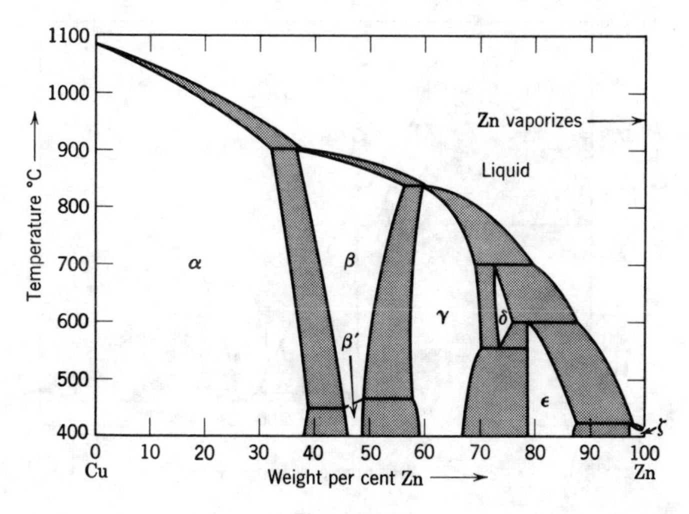

Figure VI.6

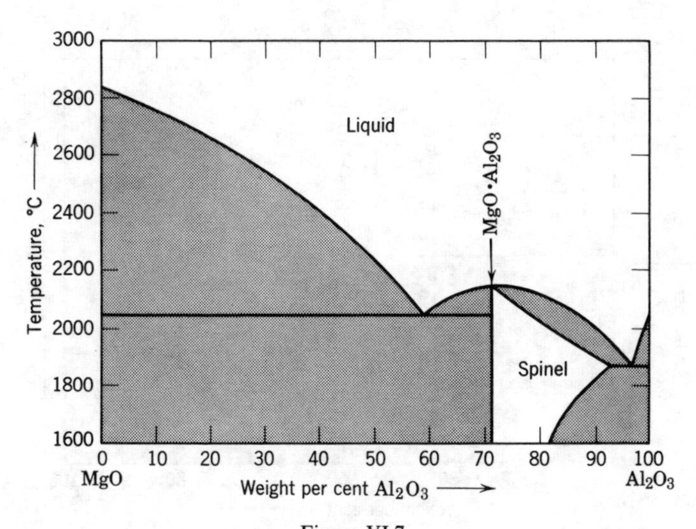

Figure VI.7

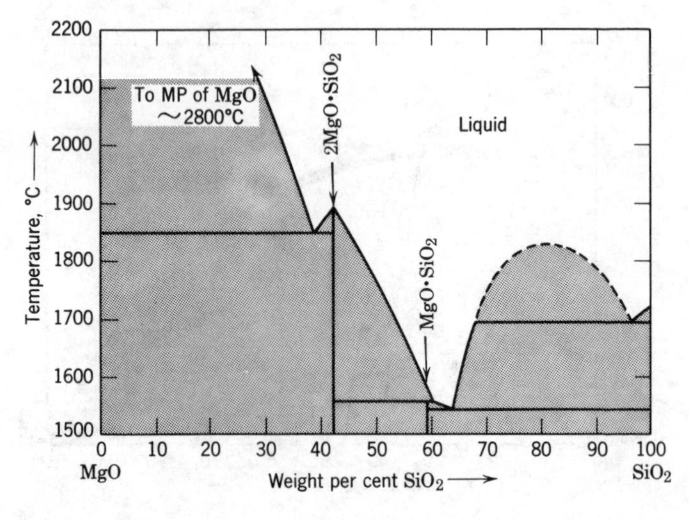

Figure VI.8

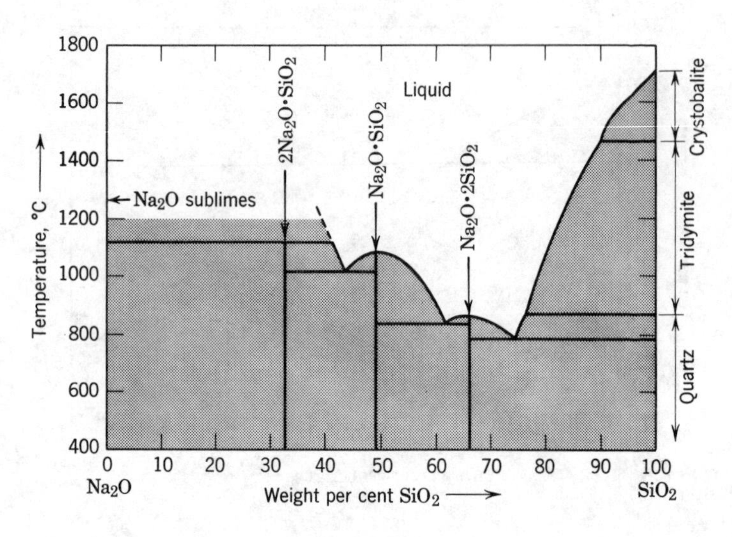

Figure VI.9

Index